Consumption and Public Life

Series Editors: **Frank Trentmann** and **Richard Wilk**

Titles include:

Mark Bevir and Frank Trentmann (*editors*)
GOVERNANCE, CITIZENS AND CONSUMERS
Agency and Resistance in Contemporary Politics

Magnus Boström and Mikael Klintman
ECO-STANDARDS, PRODUCT LABELLING AND GREEN CONSUMERISM

Jacqueline Botterill
CONSUMER CULTURE AND PERSONAL FINANCE
Money Goes to Market

Daniel Thomas Cook (*editor*)
LIVED EXPERIENCES OF PUBLIC CONSUMPTION
Encounters with Value in Marketplaces on Five Continents

Nick Couldry, Sonia Livingstone and Tim Markham
MEDIA CONSUMPTION AND PUBLIC ENGAGEMENT
Beyond the Presumption of Attention

Anne Cronin
ADVERTISING, COMMERCIAL SPACES AND THE URBAN

Jim Davies
THE EUROPEAN CONSUMER CITIZEN IN LAW AND POLICY

Jos Gamble
MULTINATIONAL RETAILERS AND CONSUMERS IN CHINA
Transferring Organizational Practices from the United Kingdom and Japan

Stephen Kline
GLOBESITY, FOOD MARKETING AND FAMILY LIFESTYLES

Eleftheria Lekakis
COFFEE ACTIVISM AND THE POLITICS OF FAIR TRADE AND ETHICAL CONSUMPTION IN THE GLOBAL NORTH
Political Consumerism and Cultural Citizenship

Nick Osbaldiston
CULTURE OF THE SLOW
Social Deceleration in an Accelerated World

Amy E. Randall
THE SOVIET DREAM WORLD OF RETAIL TRADE AND CONSUMPTION IN THE 1930s

Roberta Sassatelli
FITNESS CULTURE
Gyms and the Commercialisation of Discipline and Fun

Kate Soper, Martin Ryle and Lyn Thomas (*editors*)
THE POLITICS AND PLEASURES OF SHOPPING DIFFERENTLY
Better than Shopping

Kate Soper and Frank Trentmann (*editors*)
CITIZENSHIP AND CONSUMPTION

Yolande Strengers
SMART ENERGY TECHNOLOGIES IN EVERYDAY LIFE
Smart Utopia?

Lyn Thomas (*editor*)
RELIGION, CONSUMERISM AND SUSTAINABILITY
Paradise Lost?

Harold Wilhite
CONSUMPTION AND THE TRANSFORMATION OF EVERYDAY LIFE
A View from South India

Consumption and Public Life
Series Standing Order ISBN   978–1–4039–9983–2 Hardback
                              978–1–4039–9984–9 Paperback
(*outside North America only*)

You can receive future titles in this series as they are published by placing a standing order. Please contact your bookseller or, in case of difficulty, write to us at the address below with your name and address, the title of the series and the ISBNs quoted above.

Customer Services Department, Macmillan Distribution Ltd, Houndmills, Basingstoke, Hampshire RG21 6XS, England

---

Smart Energy Technologies in Everyday Life

Smart Utopia?

Yolande Strengers
RMIT University, Melbourne, Australia

© Yolande Strengers 2013
All rights reserved. No reproduction, copy or transmission of this publication may be made without written permission.

No portion of this publication may be reproduced, copied or transmitted save with written permission or in accordance with the provisions of the Copyright, Designs and Patents Act 1988, or under the terms of any licence permitting limited copying issued by the Copyright Licensing Agency, Saffron House, 6–10 Kirby Street, London EC1N 8TS.

Any person who does any unauthorized act in relation to this publication may be liable to criminal prosecution and civil claims for damages.

The author has asserted her right to be identified as the author of this work in accordance with the Copyright, Designs and Patents Act 1988.

First published 2013 by
PALGRAVE MACMILLAN

Palgrave Macmillan in the UK is an imprint of Macmillan Publishers Limited, registered in England, company number 785998, of Houndmills, Basingstoke, Hampshire RG21 6XS.

Palgrave Macmillan in the US is a division of St Martin's Press LLC, 175 Fifth Avenue, New York, NY 10010.

Palgrave Macmillan is the global academic imprint of the above companies and has companies and representatives throughout the world.

Palgrave® and Macmillan® are registered trademarks in the United States, the United Kingdom, Europe and other countries

ISBN: 978–1–137–26704–7

This book is printed on paper suitable for recycling and made from fully managed and sustained forest sources. Logging, pulping and manufacturing processes are expected to conform to the environmental regulations of the country of origin.

A catalogue record for this book is available from the British Library.

A catalog record for this book is available from the Library of Congress.

Contents

Preface	vi
Acknowledgements	viii
About the Author	x
List of Abbreviations	xi
1 Introducing the Smart Utopia	1

Part I

2 Imagining the Smart Utopia	17
3 Resource Man	34
4 Energy in Everyday Practice	53

Part II

5 Energy Feedback	73
6 Dynamic Pricing	94
7 Home Automation	116
8 Micro-generation	135
9 Reimagining the Smart Utopia: A Conclusion	155
Glossary	168
Notes	173
Bibliography	177
Index	199

Preface

This book is not a 'typical' analysis of smart energy technologies – if indeed there is such a thing. It therefore seems helpful to clarify from the outset the different fields of research this book intersects with and the audiences it seeks to reach. While I broadly identify myself as a social scientist, the analysis presented here departs from the traditional 'behavioural' or 'demand-side' spaces typically reserved for scholars in my field. Drawing on theories of social practice and understandings of materiality from Science and Technology Studies (STS), I seek to understand how smart energy technologies, and energy itself, are participating in and potentially transforming everyday practice.

As such, this book is deeply indebted to social practice theorists and scholars, particularly those who have been instrumental in reconceptualising the role of the consumer and consumption more broadly. I draw on this body of theory to deliberately blur the boundaries between supply and demand, and to develop a material account of how energy, and the technologies that deliver it to the home, come to matter to practice in new and intriguing ways.

However, my hope is that this book will not only hold relevance for my social science colleagues, but that it will speak to the engineers and economists who are designing, building and making the case for smart grids and meters. Towards this end this book draws, as much as possible, on research from these disciplines to reconceptualise and extend the possibilities for smart energy technologies in everyday life.

Additionally, this book seeks to speak to a third audience; namely those researchers and professionals involved in designing, testing and innovating smart devices in homes. I am referring here to design disciplines such as human–computer interaction and user-centred design, where theories of social practice and STS understandings of materiality are beginning to make in-roads. My brief encounters with these disciplines have provided much inspiration and conceptual linkages that enhance the potential for smart energy technologies to reduce or shift energy demand.

Finally, I want to make clear from the outset that this book does not constitute a report on my empirical research; although when I originally began this project I thought that it would. Indeed, one of its aims is to step outside detailed case studies and represent the smart energy project as a united and global utopian agenda. Nonetheless, this book is informed by my PhD on smart metering demand management programmes and subsequent research presented at Australian and international industry and academic conferences, and published in articles and book chapters.

More specifically, research projects conducted with my colleagues for the Australian energy industry have allowed me to develop and introduce the ideal smart energy consumer – Resource Man – to groups of (primarily male) energy industry professionals, where he has incited a laugh or two and encouraged critical reflection on what, and who, the energy industry is trying to realise with the global smart energy project.

Given the rise of smart in many domains, but most particularly in the energy sector, this disciplinary encounter is timely and necessary, and in keeping with the theme of this book, my utopian aspiration is that it will continue into the future.

Acknowledgements

Writing a book is like Latour's (1987a) depiction of a 'black box', the inner workings of which are closed for discussion and assumed to be a relatively straightforward input-output process. There are many people to whom I wish to convey my deepest appreciation for helping me open my black box, uncover its inner workings and, most importantly, close it again.

I thank those people who showed me the box and encouraged me to get in it, particularly Anitra Nelson and Zoe Sofoulis. During my time in the box, many people provided insights and critical reflection for which I am deeply grateful. In particular, I thank my RMIT University colleagues Larissa Nicholls, Kim Humphrey and Ralph Horne, as well as members of the Beyond Behaviour Change social practice theory reading group. Brief discussions with Nortje Marres and Elizabeth Shove also provided much inspiration.

Sincere thanks go to Gay Hawkins, who gave me the courage to stay in the box, and whose sharp intellect and critical insights on early and final drafts, combined with her ongoing encouragement and generosity, enabled me to enrich this project.

I am also deeply indebted to my colleague, friend and collaborator Cecily Maller, who helped me develop many of the ideas in this book and introduced me to the Golem (Chapter 7). Cecily regularly went beyond the call of collegial duty, providing calm words of encouragement and sharp insight during times when the box seemed like a very dark and lonely place.

A Vice Chancellor's research fellowship from RMIT University provided me with the opportunity to spend time in the box. I am very grateful to the university for this opportunity, as well as for the flexible working conditions that have enabled me to work intensively on this project.

The publishers at Palgrave Macmillan have assisted in unravelling many of the inner workings of the box, and I am grateful for their expertise and assistance, particularly commissioning editor Andrew James and editorial assistant Naomi Robinson. I thank the special series editors, Richard Wilk and Frank Trentmann, who provided important input during the project proposal stage. I also wish to thank the anonymous reviewers of the proposal for their feedback and Sarah Shrubb, who checked the text for language and consistency.

This project has also benefited greatly from my activities outside the box, particularly from my engagements with the Australian and international energy industries over many years, including numerous discussions with academics, policy makers and industry professionals who are grappling with many of the problems discussed in this book. I am grateful to these people for many generous and frank discussions, and hope they see this book as a positive contribution to smart energy issues.

My sincere thanks go to the hundreds of households who have been involved in my energy and water research over the past eight years. While this book does not empirically report on this work or these householders' practices and experiences, it is greatly informed by their everyday lives.

Final thanks to my very smart friends and family for indulging me with many discussions on this topic. Special thanks go to my husband, Ben Goodall, who patiently engaged with me and the box during the whole project, and provided solid walls to bounce ideas off. He also regularly reminded me that there is life outside the box, and made sure I didn't take the box too seriously.

Many thanks to all.

About the Author

Yolande Strengers is a Vice Chancellor's Research Fellow in the School of Global, Urban and Social Studies at RMIT University in Melbourne, Australia. She is based at the Centre for Urban Research, where she co-leads the Beyond Behaviour Change Research Program with Cecily Maller (http://www.rmit.edu.au/ahuri/beyondbehaviour). Yolande's research, while rooted in sociology, connects with applied research agendas in cultural studies, geography, anthropology, science and technology studies and human–computer interaction design. Her PhD, on smart metering demand management programs (2006–9), involved theoretical and empirical developments in smart energy technologies and everyday practice. Since then, her research has involved applied research projects with government departments, companies and other organisations seeking to achieve socio-technical change that improves sustainability and demand management outcomes. Yolande holds a PhD in Social Science (RMIT), a Masters in Social Science (RMIT) and a Bachelor of Arts (Monash). Prior to entering academia, she worked in the energy and sustainability sectors in communications and media management positions.

List of Abbreviations

ABC	Attitudes, Behaviour, Choice
ACHE	Adaptive Control of Home Environments
AHAM	Association of Home Appliance Manufacturers
CO_2	carbon dioxide
CPP	critical peak pricing
CPR	critical peak rebate
DIY	do it yourself
DLC	direct load control
DOE	Department of Energy (US)
ESCO	energy service company
GE	General Electric
HAN	home area network
HCI	human–computer interaction
HEM	home energy management
ICT	information and communication technology
IHD	in-home display
kWh	kilowatt hour
MIT	Massachusetts Institute of Technology
PG&E	Pacific Gas & Electric Company
PV	photovoltaic
RTP	real-time pricing
STS	science and technology studies
TOU	time-of-use (pricing)

1
Introducing the Smart Utopia

Smart stuff has captured the imaginations of governments and industries around the world. In many developed and developing countries, the 'smart' tag is attached to all manner of things, including meters, grids, homes, phones, cars, communities, cities and even nations, where it confusingly characterises both the proliferation of new information communication technologies (ICTs) and the rise of resource-efficient technological features (Berry et al. 2007). In its broadest sense, 'smart' represents an ultimate desired state across all aspects of contemporary life. It encapsulates ideals of efficiency, security and utilitarian control in a technologically mediated and enabled environment. Further, it is employed by its proponents as a means of imagining and realising social and technological progress, while simultaneously solving a range of social and environmental problems.

The 'smart' label extends to people who become smart consumers, citizens or users when they come into contact with smart technologies. Smart people not only use smart stuff and live in smart places; their reality is constituted by and through these technologies. More specifically, it is through the provision and use of information and technology that people are thought to *become smart*. This is underpinned by an understanding of human action where people act in rational ways and technologies determine particular courses of action. The technocratic, functional and efficient ideals of the smart tag extend to people who use data and technology to mediate and moderate their behaviour. In these ways, the word 'smart' is not only used to describe ICT-enabled things, cities and countries; it also constitutes a distinctive ontology in which smart technologies perform and establish a highly rational and rationalising form of social order.

In the residential energy sector, where this book is situated, the smart rhetoric has taken hold. There is a proliferation of smart metering and grid 'roll-outs' and trials in most developed and some developing countries, along with a growing smart home and smart automation industry. A *smart ontology* now underpins an international aspiration for smart grids, meters, homes and their associated technologies to revolutionise the ways in which electricity is provided and consumed. Smart meters and grids are the lynchpins of this vision, enabling an extensive range of smart energy tools and technologies that energy consumers are destined to take advantage of. Most simply, smart meters and grids are intelligent or ICT-enabled versions of their 'dumb' predecessors (mechanical electricity meters and grids). They are intended to improve the operating efficiency and security of the electricity industry through capabilities such as remote billing and real-time resource management.

The task intended for these technologies – and for those intended to use them – is nothing short of transformational. Smart meters and grids are expected to address a number of significant challenges facing the electricity sector, such as peak electricity demand,[1] distribution and transmission losses, fuel security, rising greenhouse gas emissions, fraud, and inaccurate billing (Darby 2010). They will do this by increasing energy efficiency, shifting demand to off-peak times of the day, and enabling the increased integration of renewable energy into electricity grids (Ngar-yin Mah *et al.* 2012). In some cases they are expected to allow for a complete decarbonisation of the electricity sector (Fox-Penner 2010). These are no small tasks.

In this book I argue that this global and ubiquitous vision for smart energy technologies constitutes a *Smart Utopia*, which resonates with and repackages technological utopian ideals from the past. Further, it imagines and performs a 'new' energy consumer who is intended to both realise and significantly benefit from this vision. Cast in the male-dominated industries of engineering, economics and computer science – and imagined in highly functional and masculine ways – I name this efficient, technologically enabled and rational consumer *Resource Man*.

Given the scale and scope of change intended for smart energy technologies and their consumers, the lack of interrogation of this vision is alarming. My first ambition for this book is therefore to critically document and analyse the emergence and rise of the smart ontology underpinning the Smart Utopia and the ideal consumer, Resource Man. My second ambition is to understand how this vision is encountering everyday life, and what this means for the Smart Utopia's aims of reducing and shifting energy demand. I do this by analysing the problems and potentialities of smart utopian strategies as they intersect with an alternative *ontology of everyday practice*, in which smart energy technologies, and energy itself, are entangled in everyday activities. In doing so, I reject the implication that all non-smart activity is 'dumb' or without order and intelligibility. Drawing on theories of social practice, I conceptualise energies and smart technologies as participants in the everyday practices that householders perform, where these 'materials' disrupt and potentially reorder everyday routines (Reckwitz 2002a; Shove *et al.* 2012; Warde 2005).

There are many reasons to pursue this agenda, not the least of which is because relying on any one ontology sets limits on what is real and what can be known to be real (Law 2009). In relying entirely on the smart ontology, the Smart Utopia will remain trapped in a self-reproducing performance. Resource Man will form the boundaries around who or what an energy consumer can and should be, and commitment to this characterisation will reinforce and naturalise an idealised vision of rational and efficient consumption, even if it does not eventuate as planned. More problematically, ongoing commitment to the smart ontology is likely to reproduce a series of problems that undermine the aims of the Smart Utopia, such as the emergence of energy-intensive 'smart' lifestyles featuring unprecedented levels of electrically-enabled 'pleasance', or rather new energy-intensive experiences and expectations of pleasure.

My reason for envisioning another possible future for smart energy technologies is therefore not only to highlight the gaps and holes in the smart ontology and what it potentially excludes, but to demonstrate that there are other realities which perform quite divergent possibilities for achieving – and undermining – the aims of the Smart Utopia. In the remainder of this chapter I introduce the disciplinary traditions and resources I draw on to conceptualise the role of smart energy technologies in everyday life. I conclude by briefly outlining the smart questions this book seeks to answer, and the structure and scope of my argument.

Everyday practice: conceptual tools

Interrogating the smart agenda requires an explicit acknowledgement of and departure from the theories and concepts that underpin it, particularly rational choice theory, behavioural and informational-deficit models, and an underlying commitment to linear technological transfer and substitution. Instead, it requires conceptual resources and tools that provide a nuanced understanding of the interconnected role of technology and consumption in everyday life. Here I outline four interwoven theoretical and conceptual strands this book draws on to depict an ontology of everyday practice in which all human action and social change takes place through participation in social practices.

The most important of these conceptual resources are theories of social practice. Recent iterations have revived this body of theory's relevance in studies of consumption (Røpke 2009; Shove et al. 2012; Warde 2005), where they have been put forward as an alternative to dominant paradigms that prioritise individuality or social totality. The first of these paradigms proposes that individuals and their attitudes, behaviours and choices form the basis of action, while the second contends that social structures, norms and forces act upon and control action. In studies of energy consumption, these two philosophical traditions remain the dominant means of understanding the world, and in the Smart Utopia it is the first of these that dominates understandings of social action and change.

In response – and in some ways in reaction – to this methodological individualism and social normativity, social practice theory has captured the interest of a small but growing group of enthusiasts (Gram-Hanssen 2008; Halkier et al. 2011; Røpke 2009; Shove et al. 2012; Spaargaren 2011; Strengers & Maller 2011; Warde 2005). This resurgence has been paralleled by a similar interest in social practice theory in other disciplines and domains, such as media studies (Couldry 2012; Postill 2010), geography (Everts et al. 2011), and human–computer interaction (HCI) and user-centred design (Kuijer & De Jong 2011, 2012; Pierce et al. 2011; Scott et al. 2012). Recent interpretations have attempted to make this body of theory directly relevant to studies of energy demand and smart technologies (Gram-Hanssen 2009, 2010, 2011; Røpke & Christensen 2012; Røpke *et al.* 2010; Shove 2004, 2010a). While there are significant differences between them, social practice theories are united in their view that practices, rather than individuals or their normative subjectivities, constitute and mediate social reality. What constitutes a practice is also the subject of debate, but there are a few points of agreement. For example, most agree that 'shared embodied know-how' is the foundation of practice (Schatzki 2001: 3). There is also significant agreement that objects, technologies or 'things' mediate or constitute social practices in some way (Schatzki 2001). There is much more that could be said here. However, for now I wish to put these debates and distinctions to one side, and simply say that I follow recent definitions of practice as being constellations of elements (the practice entity) that are routinely performed or enacted (practice performances) (Schatzki 1996; Shove *et al.* 2012). Leaving aside another bone of contention regarding the elements that constitute the practice entity (Schatzki *et al.* 2001), I follow Shove and colleagues' (2012) simple articulation of the elements of practice as being *meanings, skills* and *materials.*

The second conceptual strand I draw on is focused on the study of the everyday – a body of research which is strikingly absent from the Smart Utopia. Indeed, in many 'smart' studies, people are entirely absent. The field of the everyday is closely connected to, but does not necessarily follow, the theoretical traditions of social practice. De Certeau's (1984) book *The Practice of Everyday Life*, for example, is sometimes considered a key work in social practice theory, but in many ways his orientation bears very little connection to modern-day theorists such as Schatzki (1996, 2002) or Shove and colleagues (2012). Similarly, the philosophy of Lefebvre (2004), and his book *Rhythmanalysis: Space, Time and Everyday Life*, offer insights to social practice scholars seeking to understand the ordering of rhythmic routines (Shove *et al.* 2009), but arguably does not offer a distinctive theory of practice. Other scholars have studied the everyday as a site of social (and technical) reproduction and change (Macnaghten 2003; Michael 2006; Pink 2004, 2012b; Sofoulis 2005; Trentmann 2009; Wilhite 2008a). Here the term 'everyday' is broadly used to describe a suite of activities routinely enacted in the course of everyday life, but it is not only the domain of those working with theories of practice. More specifically, the term 'everyday practice' is commonly used in relation to the domestic setting. It is often used colloquially, or outside theories of social practice. However, everyday practices have also become a site of growing social enquiry by consumption scholars seeking to understand the dynamics of practices that use energy and water, such as heating (Gram-Hanssen 2010), cooling (Strengers & Maller 2011), showering (Hand *et al.* 2005), freezing (Hand & Shove 2007) or emerging ICT-enabled practices (Røpke & Christensen 2012; Røpke *et al.* 2010). Following this tradition, I use the term 'everyday' as a way to loosely qualify the site or suite of practices on which I focus my enquiry – namely those that are performed routinely in and around the home.

The third set of conceptual resources I am interested in are the understandings of materiality drawn primarily from science and technology studies (STS), which have more recently intersected with social practice theories. More specifically, STS-inspired conceptualisations of materiality take material things beyond their largely passive role in theories of material culture, where culture is often thought to be inscribed into and simply 'do its work' on society. Materiality has long been (and still is) the subject of well-documented debate in STS, and I do not wish to spend too much time here tracing these theoretical developments. Nonetheless, notable STS scholars, such as Latour, Law and Haraway, have made considerable progress in developing concepts of materiality that give non-humans agency (Latour 1987b, 2000, 2005), seek to understand the performative and provisional nature of seemingly 'hard' material objects (Law 1993, 2004) and blur the boundaries between technologies and humans (Haraway 1991).

Social practice scholars, particularly Shove and colleagues (2012; 2007), have built on these theoretical traditions to position material entities as elements of practice which are integrated into and actively constitute practices as they are performed. Rather than materials being what humans tame or domesticate and appropriate through usage, Shove *et al.* (2012: 73) emphasise that the role of materials in practice is provisional and transforming: practices and their materials are always 'on the move' in a co-dependent relationship.

Introducing the Smart Utopia 7

Despite significant intellectual resources being devoted to the role of materials in practice, little attention has been paid to the role of 'immaterial materials' such as energy (Pierce & Paulos 2010), or to big systems and infrastructures, such as smart grids, meters and microgeneration systems (Strengers & Maller 2012). Important questions remain here, such as: what role does an electricity grid play in practice? How do systems of energy provision shape the ways we cool our homes and do the laundry? Can energy itself make 'demands' on or co-shape our everyday actions? This book develops this conceptual terrain, positioning smart energy technologies and energy itself as material elements of practice. In this way I am able to consider the role smart devices and the different energies they manifest play in everyday practices, and how the realities intended for smart technologies encounter, transform and are sometimes rejected from everyday life.

A final conceptual orientation to note is the two related understandings of performativity I adopt. The first of these follows social practice theorists' understandings of practices as performances enacted by those who carry them out (Reckwitz 2002b; Schatzki 1996; Shove et al. 2012). We do not find a smart energy consumer making rational choices or automating their appliances in this conceptualisation; instead, householders perform practices of which smart energy technologies – and energy itself – are, or are not, a part. By implication, the 'carriers' of practice – which in this case are householders – are involved in enacting and constituting their reality (or realities) through the practices they participate in.

The second understanding of performativity follows the 'performative turn' in the social sciences, particularly in STS, which positions different ontologies and epistemologies as performative, that is, as performing divergent and sometimes multiple realities (Licoppe 2010). I adopt this understanding to explore the performative potentialities of smart energy technologies. This also allows me to 'counter culturally constructed categories' such as the consumer construction of Resource Man, and describe a set of practices which are intended to bring his reality and social consequences into being (Butler 2010: 147; Law 2004, 2009). These two related understandings, of practice-as-performance and practice-as-performative, are intertwined throughout this book to interrogate the realities that the Smart Utopia both seeks to perform and is performing through practice.

8 Smart Energy Technologies in Everyday Life

Smart questions

Despite forming the basis for international policy reform, the Smart Utopia, as a vision of the future, has so far failed to receive significant critical attention. An explosion of international research has evaluated and predicted the costs and benefits of smart metering and smart grids, and outlined in significant detail the anticipated role of and for the new energy consumer; but what of the vision itself? What are the assumptions, histories, politics and predictions embedded in the Smart Utopia? Who are the new energy consumers intended to realise the vision? What realities is this vision performing? And how is this vision being performed and transformed in and through everyday practice? This book pursues these questions in two parts.

Part I presents two contrasting ontologies of social order. The first is the smart ontology, which is the ontology underpinning the Smart Utopia. This constitution of reality positions social action and change as being mediated by a linear and rational model of information exchange and technical substitution. The second ontology of everyday practice positions human activity as being mediated through participation in materially constituted social practices. I do not contrast these ontologies in order to suggest that there is one 'reality' proposed by the Smart Utopian clashing against the 'reality' of everyday practice; indeed, both realities are equally 'real'. Rather, what I am interested in are the intersections *between* these realties. How do the devices and strategies designed and intended for a 'smart' reality encounter the home? And what are the performative possibilities of this smart stuff as it is integrated into, or rejected from, everyday practice?

I begin the task of critically interrogating the Smart Utopia in Chapter 2, where I argue that this vision not only encompasses the electricity industry's future, but an entire way of life. More specifically, I depict the Smart Utopia's distinctive and self-reproducing smart ontology, in which technology and data mediate and manage all human experience. Chapter 3 homes in on the ultimate energy consumer, Resource Man, who is intended to realise the Smart Utopia. Characterised as the son of the well-known Economic Man (*homo economicus*), and imagined in the masculine image of utility providers, Resource Man is a technologically interested, educated and informed resource manager of the home. My aim here is to illustrate the pervasive and narrow vision of and for this smart energy consumer, and to introduce the potential problems with this aspirational character. What concerns me most is the complete absence of lived experience from Resource Man's life – a critical gap that has the potential to completely undermine the aims of the Smart Utopia.

Chapter 4 brings everyday life back into focus, drawing on theories of social practice and concepts of materiality to depict an ontology that takes everyday activity as its focal point. In doing so I attempt to account for much of the activity that is ignored, dismissed or implicated as 'dumb' and disorderly in the Smart Utopia and the characterisation of Resource Man. Rather than positioning technologies as tools with defined uses and purposes oriented towards rational ends, I position smart energy technologies as materials of everyday practice, where they sit alongside the different meanings, materials and skills of a practice (Shove *et al.* 2012). Extending these ideas, I propose that smart energy technologies, in addition to being materials in their own right, also manifest different material energies that make 'demands' on everyday practice, and bring particular meanings and ways of handling to bear on how practices are performed (Strengers & Maller 2012).

Part II empirically investigates how the vision for the Smart Utopia is being performed through householders' everyday practices. As such it explores the intersections between these two distinct ontological realities as they encounter each other in the home. In following this agenda I focus on four illustrative smart technology strategies. Chapter 5 analyses householders' use of consumption feedback which provides near-real time and historical information on a household's consumption, greenhouse gas emissions and energy costs; Chapter 6 introduces time-based pricing, which charges higher electricity rates for peak times of the day or during critical peak 'events'; Chapter 7 investigates home automation technologies such as direct load control[2] which automate specific appliances during peak times; and Chapter 8 analyses on-site micro-generation[3] or micro-grids.[4] There are other smart strategies that I do not devote much attention to, such as the growing uptake of electric vehicles and the increasing gamification[5] of energy demand, as well as many more new ideas and innovations currently being developed.

A further focus of my analysis is on two of the problems that the Smart Utopia seeks to solve or alleviate, namely peak electricity demand and greenhouse gas emissions produced by electricity production and consumption. Importantly, these are not necessarily mutually compatible issues. Reducing average energy consumption can reduce greenhouse gas emissions while making the peaks more acute, or in industry terms it can reduce the 'load factor' of the operating plant,[6] which reduces efficiencies and increases costs. Conversely, reducing peak demand does not necessarily result in a corresponding decline in total energy usage. In many cases demand is simply shifted to other times of the day. Nonetheless, these problems are commonly discussed together in policies and reports promoting smart technologies, where a series of strategies are intended to reduce both peak demand and greenhouse gas emissions.

While my analysis is focused on four smart strategies, I am interested in what they mean for how householders do the laundry, cool their home, or coordinate their daily dinner routines. What happens to smart strategies when they are introduced into a household? Where is Resource Man in this mix? Can we find him making rational decisions on behalf of his household, and/or automating his everyday practices with technology? Where do all the other members of the household feature in this process? And what are the limitations of and possibilities for these smart strategies in transforming everyday activity and realising energy demand reductions?

These are not easy questions to answer given that the research, data and available evidence is firmly situated within the smart ontology, where problems are defined in relation to ICT deficits that require ICT inputs. Thus I draw on available evidence from a range of sources and disciplines, including energy industry research, related studies from HCI, STS, computing science and behavioural economics, as well as my own empirical research, conducted over the past eight years. In these chapters, Resource Man takes a back seat, and instead we find householders' conflicting and sometimes contradictory imaginations, integrations, interrogations and innovations of smart energy strategies in their everyday lives.

Chapter 5 begins this empirical analysis by exploring various forms of energy feedback enabled or promoted by the Smart Utopia. The provision of feedback is viewed as an essential mechanism for allowing 'consumers to make informed choices about the way they use electricity' (AEMC 2012: 4). However, I argue that the this strategy often lacks relevance in the context of everyday life, being interpreted within a limited range of energy-saving actions that ignore or overlook many practices deemed non-negotiable or necessary. This leaves other practices that use energy, but that are not readily identified with energy-saving, to move in more resource-intensive directions, thus negating the benefits of feedback over time. However, I do not conclude that feedback is irrelevant to everyday practice. Rather, I find many other forms of social, material and embodied sensory feedback that are integral to how practices are performed and transformed.

In contrast to energy feedback, the strategy of critical peak pricing (CPP) reveals the elasticity of everyday practice during short and defined periods of peak demand. Rather than viewing CPP as a rational pricing response, in Chapter 6 I argue that this strategy 'works' because it creates an 'exceptional circumstance', much like a drought, bushfire, blackout or other disturbance, which disrupts the meanings of energy in the practices that use it, and gives electricity new material qualities of scarcity and value. Instead of being a major inconvenience or disruption, exceptional circumstances are positioned here as a normal feature of everyday life. Going against the grain of smart utopian assumptions, I conclude that CPP demonstrates the increasing malleability of household routines and possibilities for policies and strategies to reorient these in less peaky directions.

Home automation technologies reveal other possibilities for smart energy technologies beyond their intended role as passive automatons of practice. In Chapter 7 I focus on the visions and meanings of control engendered by automation devices such as direct load control, smart thermostats[7] and smart appliances.[8] I argue that they may serve to legitimise the practices that are automated by positioning energy as inconsequential in those practices. Further, I find that automation technologies are implicated in 'enhancing' practice by realising an ICT-enabled and electricity-dependent smart lifestyle featuring unprecedented levels of luxury and 'pleasure' (Lutron 2012).

However, studies of technologies in different domestic settings reveal other possibilities for automation technologies. Here I find that smart appliances can play a coordinating role in practice, in which householders gain control over their everyday routines. In religious communities and studies of the fully automated home I find that householders assign control of the performance of practices to these technologies. However, like the human-like Golem of Jewish mythology, these technologies can also 'act back', reconfiguring these performances and making 'demands' on practice. Additionally, the complexity of some automation technologies can render them uncontrollable, resulting in their rejection from practice, or in the emergence of new high-tech do-it-yourself practices of installing, operating and maintaining these devices. Far from being passive bystanders, I conclude that automation technologies are enrolled in a dynamic interplay between who or *what* is in control that has implications for when, how and how much energy is consumed.

Similarly, micro-generation represents more than a simple technological switch from one way of generating and supplying power to another. In Chapter 8 I reframe electricity supply systems as energy-making practices that produce their own distinctive energies. Focusing on the micro-energies made at the household scale, I pay attention to how these 'materials' are subsequently integrated into domestic routines. This analysis allows me to ask questions about what qualities and meanings different energies bring to practice, what participation in micro energy-making practices looks and feels like, and how practices of making and using energy potentially intersect. I find that householders can liken their experiences with making energy to gardening and cooking, rather than to the ideals of resource management. Further, I demonstrate how the design, location and characteristics of energy systems can enrol and unenrol householders in energy-making practices that intersect with practices that use energy in ways that both increase and decrease energy demand.

Bringing the threads of this book together, Chapter 9 concludes by asking and responding to some speculative questions about what a reimagined Smart Utopia might look like. If not Resource Man, then who, or what, will realise the Smart Utopia's aims? How else might we imagine smart strategies to 'work' and what does this mean for the scope of strategies possible? What other disciplines and conceptual resources are required to realise and reproduce realities outside the smart ontology? And finally, what might the future look like without anything smart or utopian at all?

Finally, it is important to make clear that this book is neither utopian nor anti-utopian, but can be read as both a warning and a prophecy. I am neither 'for' nor 'against' the Smart Utopia; however, while I sympathise with its underlying ambitions, I find its methods and ability to achieve those ambitions unconvincing. I am deliberately highlighting and critiquing a pervasive vision for the future, not to completely dismiss it, but to attend to the reality it seeks to perform, as well as to what it *does not* perform, and what else it *could* perform. In doing so this book also reproduces a different reality – one grounded in the practices of everyday life. This is not without flaws of its own, and I do not wish to suggest that it is the only way in which the scope and agenda of the Smart Utopia can be reconceptualised. Rather, my ambition is for this book to become part of the wider endeavour that is needed to extend the ontological realities in which smart technologies and their associated strategies are imagined to do their work, and actually *do* 'work'. There is no 'one' way in which this occurs; multiple realities can and do exist. However, in the Smart Utopia one reality underlies and constitutes the international vision for smart energy technologies. I begin with this vision, and the smart utopians who are imagining it.

Part I

2
Imagining the Smart Utopia

In 1516 Thomas More (2005)[1] published his famous book *Utopia*, depicting an island oasis where people live harmoniously and without adversity. His fictional title, a word literally meaning 'nowhere', has entered our language to refer to an imagined perfect place or state of things. A proliferation of technological utopians emerged in the late nineteenth and early twentieth centuries, putting forward their dreams for a future in which 'progress was precisely technological progress' (Segal 1986: 119). Like More, the aim was to achieve a 'perfect society' – one free of crime, disorder, mess, chaos, dirtiness and hardship, and fuelled by 'clean, quiet, powerful electricity' (Segal 1986: 123). The irony is that these utopian visions of 'nowhere' often ended up going nowhere, never being realised, or at least not in the ways these utopians imagined.

Almost five centuries after More's vision, we find ourselves in the midst of another Utopia. Just as large-scale public infrastructure projects were previously hailed as a solution to social and political malaise, so too are smart technologies now suggested as the way to solve a range of complex and differentiated policy problems, such as climate change, peak electricity demand, energy security and rising infrastructure costs. Even though smart energy technologies are being installed in the homes of today, the vision intended for them is yet to be realised. It is timely then to pay attention to the productive work of this vision and the reality it seeks to perform.

In this chapter I contend that the current collection of smart energy technologies, policies and programmes constitute an imagined vision of the future – a Smart Utopia – with a unified and unifying smart ontology. This is a deliberate, necessarily simplifying and somewhat controversial move that I make for several reasons. First, it allows me to draw on an extensive history of imagining the future, in which I find striking similarities to the current vision for smart energy technologies, as well as significant warnings and failures. Second, it enables me to problematise the neutral and instrumental agenda of smart technology 'roll-outs' and 'deployments' in which smart meters and grids are expected to catalyse and realise prophesied social transformation. And third, it allows me to articulate and interrogate the smart ontology that underpins this Utopia, in which social transformation takes place by and through unparalleled access to ICT and quantitative data.

Not everyone would agree that there is one unified vision for smart energy technology, and indeed in many ways there is not. Not only is the Smart Utopia arguably changing as the technologies intended to realise it advance and evolve, but it is also discussed in a highly differentiated and context-specific manner, in relation to different energy issues in different countries, with different infrastructures, histories, cultures and people. Many aspects of this vision also remain hotly contested, such as whether to opt for more centrally controlled and hierarchically managed strategies and technologies, or for market-based approaches with differentiated tariffs and enhanced consumer control (Schleicher-Tappeser 2012). The mandated installation of smart meters is also the subject of debate in some communities and countries, where an anti-utopian sentiment is put forward in relation to health, privacy and safety concerns with these devices.[2] Regulation, privatisation and data ownership are other issues where there is little agreement. Indeed, there is still no common definition of what a smart meter or smart grid actually is (Sioshansi 2012).

I wish to put these distinctions and debates aside, focusing instead on a broader cross-cutting narrative that unites smart energy projects and policies around the world – a vision for the future that encompasses not only energy, but an entire way of life. I begin by introducing the Smart Utopia, the origins of this vision, and the smart utopians who are imagining it.

Imagining the Smart Utopia

Imagining and predicting the future is not something confined to fairytales; it is a necessary and productive part of the process of establishing and maintaining social order and social change, and its history dates back to astrological priesthoods and Delphic oracles (Polak 1973). More recently, emergent technologies have become 'the fuel of social imaginings' about what society should be and the potential paths it should take to steer us on a course of betterment and advancement (Sturken & Thomas 2004: 1). These visions have had a profound impact on human society, framing not only what technologies are made, but how they are marketed, used, made sense of and integrated into our everyday lives (Sturken & Thomas 2004).

Carey and Quirk (2009) cite three ways in which imagined ideas of the future have functioned in American and British life over the past two centuries. First, they have often revitalised a disillusioned or dissatisfied public, exhorting them to 'keep "faith"' in national and international goals; second, they have functioned as a 'literary prophecy', portraying the achievements of a particular ideology; and third, the development of modern technologies and information processing has allowed the public to participate in generating data trends and making choices about the future as 'a method of cleansing confusion and relieving us from human fallibilities' (Carey & Quirk 2009: 134).

A pertinent smart technology example of the productive role of future ideas is the US Department of Energy's (DOE) *Grid 2030* document, which invites its readers to 'imagine the possibilities' created by 'electricity and information flowing together in real time' (OETD 2003: i). The aims of this endeavour are explicitly utopian: 'to envision a future electric system for North America that will be considered the supreme engineering achievement of the 21st century' (OETD 2003: i). The predicted outcomes of this vision are equally large: it is to deliver 'a more prosperous, efficient, clean, secure electricity future for all Americans' (OETD 2003: i). Like past technological visions characterised by 'a world to be delivered to us by heroic engineering' (Dourish & Bell 2011: 91), the smart grid is positioned in this document as another technological masterpiece that will deliver transformative change. The 'build and supply' era of large-scale energy provision resurfaces, in which modernity is achieved through epic feats of engineering and technological progress (Kaika 2005). In this and other ways, imagining the future is as much about the past as it is about a differentiated tomorrow (Carey & Quirk 2009).

While the technology outlined in the DOE document may have changed considerably in the decade since it was published, the vision has shifted very little. Similar documents populate the websites of government and industry bodies worldwide. In the UK and Australia, 'roadmaps' and strategy documents outline how the smart grid will give consumers 'greater control and choice' over their electricity use, allowing them to reduce their bills and carbon emissions (AEMC 2011; DECC 2009: 2). Similarly, the European Commission (2006: 7) emphasises the role of 'revolutionary new technologies' in creating a 'user-centric approach for all customers'.

Reading between the lines of these smart metering and smart grid reports we find a new group of utopian thinkers – governments, utilities, consultants, engineers, economists and technologists – all working to address anticipated resource problems, while maintaining the expectations of security and reliability that are central to the energy industry's past and present. Importantly, the vision espoused in these reports is not the product of any one utopian thinker; indeed in many cases reports such as these do not acknowledge any authorship or key thought leader at all. Rather, the future imagined for smart energy technologies emerges from government, energy utility and other third party documents such as those cited above, where the utopian agenda is clearly laid out. In these and many other documents, smart technology is positioned as the ultimate utopian technology, capable of securing, improving and cleaning up the supply of electricity, as well as enabling electricity consumers to fully realise their energy management potential. Its promise and potential are described as 'enormous' and far-reaching, enabling 'improved reliability, flexibility, and power quality, as well as a reduction in peak demand and transmission costs, environmental benefits, and increased security, energy efficiency, and durability and ease' (US Department of Energy in CEA 2011: 1).

Speaking in relation to past technological utopian visions, Sturken and Thomas (2004: 3) warn that this type of utopian rhetoric generates a 'relentless optimism in new communication technologies, [which] creates an endless cycle of disappointments, since no new technology can possibly fulfill such expectations'. Indeed, although they steer societies on particular courses and directions, utopian visions are also renowned for their failure, or for their tendency to be just over the horizon, but never within our reach (Dourish & Bell 2011).

For example, the notable utopian Alvin Toffler (1980) proposed sweeping social transformations driven by technology in his 1980s book *The Third Wave*. Among other things, Toffler predicted the proliferation of the 'Electronic Cottage' (a type of smart home) and advocated ubiquitous computing more broadly, predicting that miniaturisation would make cheap and powerful minicomputers available to the world, transforming our way of life. In many ways Toffler's vision has become a reality, but in other ways it has not. Kling (1994: 155) argues that 'it is hard to take Toffler's optimistic account seriously when a large fraction of the population has trouble understanding key parts of the instruction manuals for automobiles and for commonplace home appliances like refrigerators and televisions' – a criticism that was as true in the 1980s when it was levelled at Toffler as it is now.

Closer to home, the rise of 'peaky' domestic appliances (*e.g.* appliances that create peak electricity demand), such as air-conditioners, is a pertinent example of how visions do not always go according to plan. Ackermann's (2002: 78) historical analysis of air-conditioning in the US describes how this device was imagined:

[A]s a powerful agent of the future, a marvel both scientific and magical. Its use promised to transform an uncomfortable today into a wondrous tomorrow and to restore individuals and society as a whole to physical and economic health.

While air-conditioning is now a common feature of modern homes in many nations, this vision has not always come to pass in the ways that air-conditioning's utopian promoters intended, which was for this technology to enable completely climate-controlled homes. Indeed, one of the reasons why peak electricity demand is so problematic in countries like Australia is because the vision for 'always-on' climate-controlled homes has not yet eventuated. Air-conditioned cooling is only used when householders 'need' it, with many Australians still preferring natural conditions at other times (Strengers 2010; Strengers & Maller 2011). Somewhat ironically then, smart technologies are proposed as a method of 'solving' the peak demand problem which is caused in large part by residential air-conditioning usage. One technological utopian vision is brought in to cure another. In these and other ways, technological utopias are cycled and recycled in society, performing variations on a theme, and producing a host of unintended outcomes. Visions might come to pass, but not necessarily in the ways intended.

This does not mean that the Smart Utopia is nothing more than a fantasy. On the contrary, the IT industry's unrelenting quest to find new markets is characterised by Dourish & Bell (2011: 204) as an 'explicit act of colonization'. They suggest that in time, 'the identification of everyday life as a site of computational interest becomes something of a self-fulfilling prophecy' (Dourish & Bell 2011: 204). Indeed in many ways the Smart Utopia is already a reality. Smart energy technologies exist beyond the realm of energy providers' imaginations. They are not a vague prediction of the future; millions of smart meters are already installed worldwide and smart grid trials are underway all over the world (Logica 2010). They are the subject of major policy reform and industry investment. In energy policy and industry documents, the dissemination and transformative capabilities of these smart technologies read as factual and highly probable accounts of the future. However, many promises and predictions for smart energy technology are unrealised and, as one industry analyst puts it, the vision remains an 'empty vessel that is yet to be filled with any value or significance' (Wimberly 2011: 2).

From this brief discussion we can discern that the Smart Utopia is neither a faraway fantasy nor a *fait accompli*. Rather, it might make more sense to think of the Smart Utopia as something with performative potential: that is, as something which has the potential to perform *different* or even *multiple* realities. In order to consider these possibilities, we first need to understand what it is that the Smart Utopia seeks to perform, and how and why it does this. In the discussion that follows, I suggest that the Smart Utopia aims to perform a way of life mediated by and through technology (or, more specifically, ICT) and quantifiable data. The remainder of this chapter is devoted to illustrating how ICT and data constitute a smart ontology which underpins the Smart Utopia. This ontology is grounded in the energy industry's engineering and economic roots and extends into the home, where it positions householders as micro-resource managers who use data and ICT to mediate and manage social action and change. I begin this analysis by considering the ways in which technology is enrolled in realising the Smart Utopia.

Technology

It may seem incredibly obvious to point out that technology, or more specifically ICT, is central to the Smart Utopia; indeed the vision is never discussed without reference to some form of smart technology. However, what is rarely discussed are the ways in which this vision uses ideas and ideals of technology to characterise the world and all action within it. Technology, rather than being just a material feature of the Smart Utopia, forms part of the foundation of the reality it constructs and reproduces.

The Smart Utopia proposes a world in which social disharmony and environmental problems are eradicated through new technology, without compromising current ways of life. Technologies are employed in a number of ways to achieve this vision. First, they promise rational, efficient and ordered control of people and their environments; second, they politically position and assign responsibilities for complex environmental problems to individual consumers; and third, they seamlessly manage the home environment while maintaining or enhancing current lifestyle expectations. In this section I take stock of these different framings of technology in the Smart Utopia, drawing on their histories and speculating about their possible futures.

The intended role of smart technology in achieving efficiency, rationality and order has strong synergies with Ellul's (1976: 5) concept of technique, which 'constructs the kind of world the machine needs and introduces order...It clarifies, arranges, and rationalizes...It is efficient and brings efficiency to everything.' The 'machine', in this case, is not only the smart technologies of the home, but also their connections to centralised grids and large-scale metering systems. Through technique, the 'problem' is defined as a technical one in need of a technical solution discernable through scientific quantifiable methods. A single technology naturally asserts itself as the 'one best way' to solve a range of social, technical and environmental problems: 'its results are calculated, measured, obvious, and indisputable' (Ellul 1976: 79). Social scientists who are unable to 'talk numbers' are excluded from the process of defining the problem and identifying potential solutions, and instead it is only 'the specialist' who 'is able to carry out the calculations that demonstrate the superiority of the means chosen over all others' (Ellul 1976: 21). In this case, the chosen technology is a suite of smart stuff – meters, grids, homes and automation technologies intended to solve a host of energy problems.

Technique also characterises the ways in which smart technologies are thought to 'roll out' into the world in an orderly and efficient manner, where they are able to do their work. Like other technologies, they are positioned as 'neutral tools that function for us as detached, disinterested and unquestionably loyal servants' (Davison 2004: 86). They are 'deployed' into society, where they exert and enable a sense of control, and promise that our resource and environmental problems can and will be solved. This is a typical feature of technological visions, which view change as a simple process of technological substitution (dumb meters to smart meters) and treat the process of embedding new technologies in society as unproblematic (Geels & Smit 2000). This embodies not only Ellul's meaning of 'technique', but resonates with the 'techno-economic optimism' central to capitalist societies, where eco-efficiency is promoted as a way of curtailing the impacts of the growth paradigm, but at the same time subtly reinforces it (Davison 2001: 22).

A second point to note about technology is that, far from being a neutral tool, *'technology is itself a political phenomenon'* (Winner 1977: 323; emphasis in original), encapsulating and embodying social and political ideas, and being employed to achieve specific political aims. Smart energy technologies are part of a broader political ideology where responsibility for national and global social and environmental problems is shifted to the individual, and where market systems are delegated the task of regulation previously managed by states (Beck & Beck-Gernsheim 2002). The meter, which is already a powerful form of social control (Akrich 1992), acts to enable new forms of participation in markets and broader social and environmental problems, bringing public issues and responsibilities into the private domain of the home (Marres 2012a). 'Technique' infuses the nature of this relationship, positioning energy consumers as inputs into the technological system, where they are viewed as 'end users' or more broadly as 'demand'. Householders are reformulated as technicians, who require 'new technical instruments' such as in-home displays (IHDs) and smart appliances, which are 'able to mediate between man and his new technical milieu' (Ellul 1976: 84).

A third way in which technology is employed in the Smart Utopia is as seamless integration, as something that quietly pervades every corner of the home. 'Home area networks', 'home automation', 'home energy management systems', 'house-brain' or the 'homebus' embody this representation of technology, where a central controlling system intelligently and efficiently manages and regulates a home's energy usage (and other activities) on an occupant's behalf (Berg 1994: 171). In this use of technology, the 'environment' refers to one's surroundings, which are 'minutely managed', but where 'signs of this management are discreetly concealed' (Berry et al. 2007: 240). Similar ideals are embodied in smart grids, where technological 'agents' are anticipated to perform many daily domestic activities on behalf of home occupants (Ramchurn et al. 2012). Oksanen-Sarela and Pantzar (2001: 212) refer to this assignment of agency to technology as a process of 'cultural determinism', whereby it is 'natural' to view technologies as replacing people in everyday life contexts. In this way, technology is unproblematically brought in to do societal work.

A key feature of this seamless integration is the achievement of modernity and efficiency. This ideology can be traced back to early 1930s' visions of the 'homes of tomorrow',[3] in which efficiency was presented alongside unprecedented levels of luxury, relaxation and indulgence, with excessive energy consumption clearly on display. In these futuristic scenarios, efficiency did not relate to energy, but rather to domestic efficiency – it promised all the benefits of modern living with less effort from householders. For example, Horrigan (1986: 154) illustrates how Westinghouse's display-style 'Home of Tomorrow' celebrated its achievement of an electrical load equivalent to that of 30 average homes, 'ready to do the work of 864 servants with the flip of a switch'. The home featured 'air conditioning, an electric garage-door opener, automatic sliding doors, an electric laundry, 21 separate kitchen appliances, burglar alarms, 140 electric outlets, and 320 lights' (Horrigan 1986: 154) – luxurious even by today's standards.

Fast forward to the 2010s and the rhetoric of home automation and smart homes is incredibly similar. However, now the focus has shifted to include energy efficiency as well as domestic efficiency. Just as past technological utopians promulgated a vision of the future dependent on what were then not considered excessive amounts of electricity required to power every conceivable appliance (Segal 1986), the smart energy visionaries of today are assigning technology the task of reducing it. As well as controlling entertainment appliances (televisions, music players, recorders, computer games, home computers), safety and security features, communication within and outside the home and the environment (temperature and air pollution), visions of the smart home now include the efficient control of energy, water and waste (Berg 1994). However, they still pursue the same or even more luxurious standards of living as early smart home visions, with technology positioned as the means to achieve unparalleled levels of luxury.

For example, an Australian smart grid demonstration home report states that the aim was to 'demonstrate efficient use of electricity and water *without compromising modern lifestyle*' (Ausgrid 2012: 1, emphasis added). Other modern smart appliance and automation providers, such as the home automation company Lutron, resonate more with Westinghouse's ambition of explicitly *improving* modern lifestyle, by inviting their customers to 'experience the essence of pleasance™' (Lutron 2012). In Lutron's advertising material, energy efficiency sits alongside ambience and temperature settings and enhanced security to produce a space in which one can find 'comfort, romance and peace of mind – a place where you experience pleasance' (Lutron 2012: 1). Similar utopian visions have existed throughout history: for example, during the industrial revolution of the home the household technologies meant to unburden women from domestic duties actually served to increase cleanliness expectations and women's workload (Forty 1986; Schwartz Cowan 1989). Likewise, rather than being neutral slaves, smart technologies may reinforce and increase expectations for new or technologically mediated forms of technology or climate control that are central to the ideals of integrated energy management.

In summary, technology is central to the Smart Utopia in three complementary and sometimes contradictory ways. It functions as a tool to unproblematically bring rationality, efficiency and order to the energy system and all those who use it. It is employed as a political device intended to shift responsibilities for social and environmental problems from governments and utilities to householders. And it functions as a means of seamlessly bringing ideals of efficiency and luxury to the home, in which technology takes care of and enhances a range of domestic practices. From this brief discussion we can see that technology is by no means a neutral bystander in the Smart Utopia. It is enrolled in performing a reality in which people-who-use-energy take responsibility for energy management through rational and efficient means, while simultaneously assigning this responsibility to technology, which will take care of energy issues on their behalf and in doing so, maintain or enhance their experiences of pleasure.

Data

Just as the Smart Utopia positions technology as a remedy to solve a variety of social malaises, so too does it position data as a means to realise similar social transformation. The growth of ICTs has led to unprecedented quantities of data in the energy sector – and many other sectors – which is sometimes referred to as Big Data (Mayer-Schönberger & Cukier 2013). Simultaneously, there has been a corresponding growth in the field of 'data mining', which seeks to make sense of what these data mean (Han & Kamber 2006). Energy data are nearly always numerical, quantifiable or calculable in nature, and they are central to every aspect of the Smart Utopia in four key ways.

First, data are used to justify the costs and benefits of smart metering and smart grid projects and policies. A critical feature of these data is that they are about data. They attempt to quantify the costs and benefits of capturing data through new smart tools, and of using these data in a number of different ways to understand consumers and their consumption or to provide data to consumers in more sophisticated ways. Here data have performative consequences, being used to generate a 'literary prophecy' (Carey & Quirk 2009: 134) about a future characterised by data. Data are drawn on by smart energy advocates as a 'strategic resource' (Geels & Smit 2000: 882), where they are used to set expectations and promises about smart energy technologies. Engineers and economists are the specialist experts who collect and prepare these data. Social scientists, where consulted, are brought in at the end of this process, or to address specific aspects of an analysis, such as to provide data on how consumers might react to smart technology, or how to best encourage them to use it. Further, social scientists who can 'speak data' (particularly Big Data) are encouraged, such as those able to conduct the large-scale quantitative surveys with consumers that represent them as segments, percentages and aggregate figures. In these and other ways, data are used to create a vision of the world that is governed and ordered by data.

Second, data constitute the basis of the relationship between the providers and consumers of energy. Utilities and governments are positioned both discursively and strategically as data-providing advisers that are there to meet the preferences and serve the needs of smart energy consumers. Simultaneously, consumers are positioned as data-hungry decision-makers, who make rational choices about how to use energy in their homes based on this information. Data are delivered in a number of ways: for example, as consumption information delivered through website portals and IHDs, or through the provision and management of time-based pricing signals (charging different amounts for usage at different times of the day). Data can also be delivered to household technologies, such as washing machines and dishwashers, which can be programmed to come on at specific times of the day. Data, and the technologies that mediate it, are thereby the means through which providers and consumers of electricity interact. Through the provision and management of data (and ICT), the smart energy consumer is asked to participate in the Smart Utopia.

A third data-related feature of the Smart Utopia is their role in enabling consumers to realise the utopian vision. Data are collected through smart meters at half-hourly or more regular time intervals, analysed by electricity utilities to manage the system and discern demand profiles, and provided back to householders via IHDs or web portals as 'actionable data' (Fox & Gohn 2011: 3). Providing these data to consumers is put forward by governments and utilities as the mechanism by which many of the Smart Utopia's aims, such as reducing household bills, shifting peak demand, and alleviating greenhouse gas emissions will be achieved (see, for example, AEMC 2011; OSTP 2012). Theories of rational choice and information exchange are central here: data are thought to translate into concrete actions that consumers can and will take (such as measures to reduce their energy bill). In this way, data, together with technology, are positioned as the basis of human action and social change.

Data are also the means of understanding and *managing* energy consumers through intensive and ongoing market research and segmentation. Most recently, these data can be found in extensive research carried out to understand the inner workings of the 'new energy consumer' (Accenture 2010, 2011, 2012a; Zpryme 2011; see Chapter 3). The aim is to capture sophisticated 'analytics' about energy consumers, in order to understand and target them. This counting and collection leads to new forms of classification and control, because categories need to be invented into which people can conveniently fall and be subsequently monitored and targeted (Hacking 1990). These ideas are not new; they build on market research techniques dating back to at least the 1950s, where data were positioned as a means of both conducting consumer surveillance and disciplining and controlling consumption (Miller & Rose 1997). Critical differences now are the sheer amount of data made available through smart technologies, the computer power available to analyse them, and the speed with which these data can be manufactured into 'actionable insights' for energy consumers (Accenture 2012a; Zwick & Denegri Knott 2009).

These intensive ways of knowing consumers are performative (Law 2009). Like other forms of classification that have 'profoundly transform[ed] what we choose to do, who we try to be, and what we think of ourselves', new ways of collecting and segmenting consumers essentially equates to new methods of 'making up people' (Hacking 1990: 3, 6). Similarly, Zwick and Denegri Knott (2009: 241) argue that customer databases not only represent an irresistible epistemological position, or way of knowing consumers, but also 'manufacture consumers ontologically' by transforming 'action and inaction, movement and inertia, indeed all life' into an 'information commodity'. Van Vliet (2006: 311) goes further, observing that in an effort to know all social action through data, consumers may actually become data, being positioned as 'a series of half-hourly or night-and-day loads who are switched on and off to match certain network capacities, or as a series of hot and cold spots on infrastructure networks'.

An example of how this occurs in the Smart Utopia comes from a Smart Grid Consumer Collaborative (SGCC 2012: 16) report, which finds that '67% of consumers definitely (28%) or probably (39%) would participate in a Smart Meter Data Energy Management program'. In this and other reported surveys, consumers are asked about data, encouraged to show interest in data, and – unsurprisingly – state that they are interested in data. Questions that don't relate to data management, acquisition, or data-processing are often not asked. The risk is that consumers begin to equate 'energy management' with 'data management', rather than with the ways in which energy is consumed during the course of their day-to-day practices.

These four ways in which data is intended to function in the Smart Utopia are of course inter-related: 'hard' cost-benefit data are used to justify the large-scale 'roll-out' of smart energy technologies; data forms the basis of the relationship between providers and consumers of energy; the collection and provision of data to energy consumers enrols them in new forms of social action and change; and consumers are known, understood and targeted by the collection of data about their interest in energy data. Taking Hacking's (1982, 1990) 'avalanche of printed numbers' to a new extreme, the Smart Utopia reproduces a smart ontology in which data, along with technology, constitutes reality. This ontology has implications not only for energy consumption, but for all social life. More specifically, it has significant ramifications for how people who consume energy are positioned and understood, and for how the Smart Utopia seeks to relate to and transform them.

The smart energy consumer

On the one hand, smart energy consumers or smart home occupants are often noticeable by their absence, particularly from productive spaces like kitchens or laundries (Bell & Kaye 2002). If you look inside a promotional smart home, you are likely to find a heterosexual nuclear family living a happy, relaxing and trouble-free life with every modern convenience and luxury they could wish for (Dourish & Bell 2011). They are *unproblematic* consumers, and therefore deemed unworthy of our time and attention; we do not find disorder or chaos in these visions of the smart home (Berg 1994; Oksanen-Sarela & Pantzar 2001; Wyche *et al.* 2006: 37). The smart home is presented as a secure site of indulgence, entertainment and relaxation (Berry *et al.* 2007). Occupants, insofar as they feature at all, do so as largely autonomous, homogeneous and passive agents (Dourish & Bell 2011).

Extending these ideas of human absence that are central to smart home visions, many believe that human passivity is the only way to realise the Smart Utopia; that is, with technologies acting on behalf of people, involving minimal or no involvement by home occupants. Smart technologies such as direct load control (DLC) and set-and-forget or prices-to-devices smart appliances[4] (Hamilton *et al.* 2012; see Chapter 7) encapsulate this view that technology can take control of energy management problems on behalf of consumers. This is an appealing notion for many electricity distributors and transmitters (particularly engineers), not least because it maintains the traditional 'predict and provide' logic (Guy & Marvin 1996) of the electricity industry by keeping utilities firmly in control. It is also alluring for utilities that often do not have direct access to consumers but are required (by regulation) or at least aspire to provide reliable and sustained load at all times, regardless of what consumers do or how much electricity they use. Distributors can be blamed or even financially penalised when they are not able to provide reliable power, even when it is arguably through no 'fault' of their own (Strengers 2008). The further appeal of this passive position is that some smart energy experts simply do not agree that consumers will become motivated to manage their consumption during peak demand (Berst 2012). The simple solution is for utility providers to manage energy consumption on consumers' behalf using technology and data as mediators in this process by, for example, automatically turning appliances off or down during periods of peak demand.

Passive understandings of smart energy consumers sit alongside a second active role intended for them, whereby consumers *take control* of technology, energy and their environment. The Massachusetts Institute of Technology's (MIT) Home of the Future Consortium provides an example of this active vision. These researchers argue 'that the home of most value in the future will not use technology primarily to automatically control the environment but instead will *help its occupants learn how to control the environment on their own*' (Intille 2002: 76, emphasis added). Assisting occupants to learn how to take control, MIT's consortium continues, involves designing and constructing a living laboratory where 'our pervasive technologies... empower people with information that helps them make decisions' (Intille 2002: 77).

A similar vision for consumers pervades smart grid and metering reports, where these technologies are intended to 'empower consumers' (OSTP 2012) to 'take control' (CEA 2011) of their consumption and make 'informed choices' (AEMC 2011) about how they use energy. Resonating with Ellul's depiction of technique, the empowered and informed consumer is imagined as an extension of technology who is constantly seeking self-improvement through self-monitoring and control (Oksanen-Sarela & Pantzar 2001). Efficiency extends from machines to people who 'actively manag[e] their life through technology' (Oksanen-Sarela & Pantzar 2001: 204). This vision encapsulates modernist approaches to energy and sustainability problems, and relies heavily on the market (pricing signals and informed consumers) to ensure that necessary change is achieved (Brynjarsdottir et al. 2012). As such, it is often a feature of liberalised energy utilities in a competitive energy market (Schleicher-Tappeser 2012).

Smart energy technologies are thus intended to be both 'rational and rationalizing' (Dourish & Bell 2011: 165–6); they can make rational decisions on behalf of occupants as well as enable occupants to make rational decisions. Householders are represented as both active, present and informed consumers who are in control of their energy consumption, and passive, absent and disengaged consumers who assign control of their energy consumption to technology. While it is often argued that some consumers will only be interested in either one of these roles intended for them in the Smart Utopia, the ultimate smart consumer, Resource Man, is interested in both (see Chapter 3). Like a system engineer, he is involved in actively managing his consumption through a range of resource management tools while also passively assigning this management to smart technologies.

Harper-Slaboszewicz et al. (2012: 34) encapsulate 'the goal' of Resource Man, which 'isn't to move utilities into our living room – rather it's to allow consumers to take advantage of some of the same technologies utilities are finding useful in smart metering and monitoring/managing the distribution grid'. In other words, the aim is to transfer energy utilities' values, knowledge, expertise and technologies into the heart of the home. Thus, in a similar vein to the 'scholastic bias' academics project towards the people they study (Bourdieu 2005: 45), Resource Man represents the energy industry's 'resource bias' projected onto energy consumers. In this way we can see how the smart ontology extends from utility providers to the home, where it constructs householders as what Sofoulis (2011: 805) refers to as 'Mini-Me' versions of their utility providers.[5]

The utopian aspiration is for householders to act as micro-resource managers (Strengers 2011b) by making use of a range of technology and data-related tools.

This chapter has introduced the Smart Utopia as a unified vision of the future, underpinned by a smart ontology in which ICT and data constitute a social reality. By reinventing and reinvigorating past technological utopian and political ideals, the continual 'work' of this vision is a 'seductive description of social change driven by technology' (Kling 1994: 154). In the following chapter, I turn our attention to the new breed of energy consumers intended to realise the Smart Utopia, focusing specifically on the ultimate imagined consumer – Resource Man. More specifically, I investigate how Resource Man is being performed, resisted and contested. In doing so, I start to identify some of the cracks in Resource Man's conceptualisation, not least the complete absence of everyday life.

3
Resource Man

> Downstream, for end users, the impacts of the Smart Grid are potentially profound. Customers will face electric prices that vary within each day, and they will have far more information and control over their power use and costs. With software simple enough to run on a cell phone, they'll monitor the energy used by several appliances linked to their home network, controlling them immediately or programming them to react to prices. With the touch of a button you will be able to programme your air conditioner to turn off fifteen minutes out of every hour when hourly electric prices exceed a certain set-point. Yes, you'll be a little warmer, but you'll also save good money. And for the majority who don't want more complex power, appliances will all come preprogrammed so users can connect them seamlessly at factory default settings. (Fox-Penner 2010: 35–6)

Fox-Penner eloquently sums up the dominant vision for smart energy consumers in the quote above. They are located *downstream*, on the *demand-side* of the supply chain, where a range of choices are provided to suit their specific needs. This conceptualisation is now underpinned by an internationally consistent suite of methods, theories and assumptions, held together by the cumulative weight of ongoing, intensive consumer research. The ultimate energy consumer emerges from these reports as a rational and rationalising Resource Man. He is imagined in the image of his utopian masterminds – engineers, economists and behavioural scientists – and is positioned as an efficient and well-informed micro-resource manager who exercises control and choice over his consumption and energy options. In this way, Resource Man embodies *technique* in all his actions (Ellul 1976; see Chapter 2), by choosing a range of technological and data-mediated tools to suit his unique lifestyle.

Like the Smart Utopia itself, I represent this new energy consumer as a uniform conceptualisation, even though he is often represented as a series of highly differentiated consumer segments with different needs and preferences (Accenture 2011; SGCC 2012). A key aim of this chapter, and indeed this book, is to demonstrate that Resource Man is not a socially factual category, nor does he necessarily 'exist' as imagined, nor is he even 'new'; indeed, he bears striking resemblance to the imagined recipients and residents of past technological visions (Oksanen-Sarela & Pantzar 2001). Nonetheless, Resource Man is incredibly important because consumer conceptualisations are productive, not only representing and understanding consumers, but also producing and creating them (Trentmann 2006).

This chapter begins by introducing Resource Man more formally, outlining his ancestral origins and his gendered orientation. I continue by demonstrating how segmentation analyses and extensive consumer surveys give the illusion of diversity and differentiation, while representing consumers as socially factual categories. It is through this research that Resource Man not only comes to be known, but is also made real. Extensive efforts are now underway to transform more energy consumers into Resource Man, with varying degrees of 'success', some of which are discussed in the second half of this chapter. Efforts are also being made to engage Resource Man in new partnerships and collaborations. However, these rarely go beyond positioning energy consumers as 'learners' who are on their way to becoming expert micro-resource managers. I conclude that Resource Man represents an incredibly narrow vision of social action and change.

This chapter is informed by an impressive body of international consumer research conducted by or for energy utilities, governments, technology providers and behavioural economists and psychologists. In total, over 50 publicly available international smart metering and grid consumer reports or report summaries were reviewed for this chapter, encapsulating research conducted over the last decade with over 100,000 residential (and small business) energy consumers. It is in these reports that we find the new energy consumer – Resource Man.

Introducing Resource Man

Resource Man, or *homo facultas*, is my name for the gendered, technologically minded, information-oriented and economically rational consumer of the Smart Utopia. He is not new and nor is he my invention; rather, he is a fusion of old ideas packaged in new ways (see Chapter 2). In his ultimate imagined state, Resource Man is interested in his own energy data, understands it, and wants to use it to change the way he uses this resource. He is the ideal and idealised individual consumer of energy, and his aim is total control and choice over his use of energy so that it is operating as efficiently as possible, in a way that suits his lifestyle. For these tasks he will need data, education (about energy), different demand management options, and new enabling technologies that will allow him to transform his home into a resource control station. This new energy consumer is absolutely essential to the Smart Utopia, where he is expected to 'unlock the vast potential of the smart grid' (CEA 2011: i).

Of course Resource Man does not always go by this name. Zpryme Smart Grid Insights (Zpryme 2011) refer to him as the 'new energy consumer', which they suggest is most likely to be a 27 to 35-year-old technologically immersed, energy-literate male with a college degree or higher, and an average household income of US$70,000–$100,000. This energy consumer will pay the price for renewable sources, monitor his electricity usage on a daily basis, and drive mainstream adoption of electric vehicles, smart devices and residential solar and wind systems. This consumer currently represents 11–13 per cent of the US adult population, but according to Zpryme that figure will reach 17 per cent by 2015. While the IBM Institute for Business Value does not mention gender, they largely support Zpryme's conclusions, finding that the first wave of 'information-hungry' and 'technology-savvy' consumers are 25–35 years old, but that the Millennial Generation (18–24) are the real resource men (and women) of the future (Valocchi *et al.* 2009: 10).

At the most fundamental level, Resource Man is characterised by his ability and willingness to take control of his consumption and make individual choices about it. Control is defined in two ways: first, Resource Man can assign control of his energy consumption to technologies or utilities. Advocates of set-and-forget (Harper-Slaboszewicz *et al.* 2012: 394) or 'cruise control' (Berst 2012) technologies support this aspect of the vision, whereby smart appliances and home area networks act on Resource Man's behalf, controlling and shifting his consumption in response to price signals and other market-based data. Conversely, and often concurrently, control is also defined as Resource Man being empowered to 'take control' of his energy consumption through pricing signals and daily or real-time consumption data, which enable him to make rational and informed decisions. Choice is defined in relation to these two conceptualisations of control. In practical terms, this might involve choosing a particularly pricing tariff, home energy management system or any other programme or product. Greater choice is generally assumed to improve Resource Man's wellbeing and resource-saving capacity, although some analysts warn that too much choice can be confusing (Ipsos MORI 2012). Importantly, in order to make the 'right choices' (OL 2009: 11), Resource Man requires expert energy information, knowledge and tools.

Not all consumers are expected to attain Resource Man status in the fully realised Smart Utopia, with some analysts finding that 'a significant subset of the population [over 30 per cent]...will not pay attention to the choices offered and [will] passively continue the behavior patterns to which they are accustomed' (Valocchi & Juliano 2012: 9). However, while Resource Man may not exist everywhere, he can be located or pinned down in very specific ways. Like the consumer conceptualisations that have come before him, he can found at the *end* of the supply chain, where he acts alone and autonomously, individually managing and consuming energy. It is here, on the 'demand side' (Accenture 2010: 4) of the meter or at 'the bottom of the system' (Schleicher-Tappeser 2012: 29) that consumers are 'plugged in' (Valocchi *et al.* 2007) to smart technologies, and where their role as 'end users', 'customers' or 'consumers' of energy services begins and ends. It is also here, in the home, that Resource Man assumes his hierarchical and somewhat traditional role as the Resource Man of the house.

I refer to Resource Man as a male not because he is always directly identified as one, but because he is cast in the image of the male-dominated industries of engineering, economics and computer science,[1] and because visions of him exclude most household labour, which is still predominantly carried out by women. Berg (1994) makes a similar point in relation to past smart home visions, where she notes the masculine orientations of technological change and the absence of housework. Further, the Resource Man vision unsurprisingly captures the attention of men more than women, resonating more broadly with Caplan's (2001) finding that 'maleness' is one of the key attributes of people who think like economists. Closer to the issue at hand, American consumer research finds that men are more interested in technological solutions for their energy use, and in monitoring and managing their usage through the latest personal electronics (Accenture 2011). Resource Woman might still exist, but I use the gendered term to highlight the masculine orientation of this new consumer category and construct. I also primarily refer to him in the individual, for while there are many resource men, they are largely thought to be operating in isolation from one another.

Resource Man has many relatives and ancestors. Perhaps most obviously, he can be imagined as the son of Economic Man, or *homo economicus*, differentiated by his access to more advanced and sophisticated technology and data-enabled tools than his economically minded father. He is a close cousin of *homo optionis* (Choice Man), for whom all aspects of life are 'decidable down to the small print' (Beck & Beck-Gernsheim 2002: 5), reflecting Resource Man's extensive range of technology and data options along with his ability to choose an energy retailer to provide them. Other relatives include Average or Normal Man, *l'homme moyen* – a term that Hacking (1990: 169) describes as 'one of the most powerful ideological tools of the twentieth century'. The term is reinvented here to embody a new vision of normality, in which it is 'normal' to manage resources and everyday life through technology and data. Another important ancestor is *homo faber* or Tool Man, referring to Resource Man's recent evolutionary progression towards new ICT 'tools'. Further down the family tree, Resource Man is related to the 'entrepreneurial self' (Petersen & Lupton 1996) and the 'citizen-consumer', who are part of a neo-liberal political discourse where responsibility for a range of problems is shifted from the state onto individuals, and where energy is positioned as a commodity in a market system which relies on the actions and choices of individual consumers, who are controlled and coerced through various forms of information and heteroregulation (Hinchliffe 1996).

Taken together, Resource Man is a new formulation of old ideas and pre-existing gender biases, tightly packaged together. It is in and through consumer research and subsequent consumer programmes, products and packages that Resource Man is known, made and performed as a new consumer category. The remainder of this chapter is devoted to identifying the ways in which this is done.

Knowing Resource Man

I have so far suggested that Resource Man is a unified vision of the new energy consumer. However, this is not the way in which he is studied, represented or commonly known. Indeed, one of the paradoxes of this consumer category is that it performs an illusion of differentiation. This is frequently articulated in the growing recognition that there is not 'one best way' or a 'one size fits all' approach to understanding or engaging with energy consumers. A common view is that different representations of consumers are required with tailored value propositions, products and packages, because 'consumers are complex individuals with distinct and often idiosyncratic needs and requirements' (Accenture 2011: 29). The tactics and methods of representing these different groupings create an illusion of diversity, which co-exists with the pursuit of a penultimate or universal consumer 'solution', such as the continuing search for the elusive 'killer application' that will 'capture consumers' imaginations and motivate change in energy behavior' (Accenture 2012b: 30) Differentiation is established through market research, in which consumers are divided and categorised into demographic or psychographic segments, where different 'inputs', in the form of information, pricing programmes, smart devices or other demand management products and programmes, are targeted towards specific groups of consumers to achieve desired behavioural 'outputs'.

Segments are typically defined in relation to the cost of energy (money-minded strivers, cost-sensitives), attitudes towards energy (energy stalwarts, energy epicures), concern with energy's impact on the environment (green boomers, eco-rationals), age (senior savers and young families), interest in energy products and technologies (big toys, big spenders, tech-savvys), effort put into managing energy consumption (proactives, indifferents), and relationships with energy providers (lacking trust, traditionalists) (Accenture 2010, 2011; SGA 2011; SGCC 2012; Valocchi *et al.* 2009). Part of the appeal of segmentation is the impressive and relatable categories market researchers develop (surely we all know a senior saver?). This allows energy providers to not only 'speak human' but 'speak consumer', enabling them to know and understand their consumers in ways that can be translated into 'actionable insights' (Accenture 2012a).

Despite the façade of diversity, different segments still maintain and reproduce a cohesive Resource Man vision in which humans are defined in relation to the ICT- and data-enabled energy products, services and relationships they are individually willing to buy, use or otherwise engage in. For example, Accenture's (2010, 2011, 2012a) extensive research with over 30,000 international energy consumers frames householders' relationship to energy solely in terms of their role as consumers of it. Within this framework they are asked questions about their values related to energy, various energy and non-energy related products they might like, service perceptions of their energy providers, and choice-based questions about energy product and service packages. More specifically, energy is consistently framed as a commodity (something which consumers purchase), a resource (kilowatt hour) or an impact (on the environment), rather than something which is consumed throughout the course of everyday living or domestic activities, such as doing the laundry or cooking dinner. Unsurprisingly, then, a remarkably consistent conclusion from consumer research of this kind is that the 'ultimate motivator is electricity bill savings or incentives' (Jelly 2008: 66), in combination with a range of 'other factors' such as 'doing the right thing' for the environment, and 'including more energy control/management' (SGCC 2012: 12).

In these ways, consumer research has both ontological and epistemological effects (Law 2009). Ontologically, it represents the world as a place in which people can be divided and grouped into discrete bundles based on their attitudes, behaviours and values about energy as a resource, commodity or impact to be consumed, managed or saved. Further, it is (only) through these relationships and roles that consumers are thought to relate to energy and its consumption. Epistemologically, consumer research reproduces a way of counting, measuring and targeting strategies towards these individual attitudes, behaviours and values within discrete segments. Like concepts of the average consumer, which Sofoulis (2011: 806) describes as 'a simplistic reduction that blocks culturally intelligent appreciations of diversity, contradiction, ambiguity and multiplicity', customer segmentation reduces human action to a series of relations and relationships centred on and about energy, while giving the illusion of social and cultural complexity. When consumers are subsequently 'targeted' with energy products and choices tailored to their segment, these concepts and categories may have self-actualising effects, by implying that consumers *ought* to relate to energy in these ways. This generates what Hacking (1996: 60) describes as a 'looping effect' whereby 'the classifications and our knowledge interact with the people classified, who often change or modify their behavior simply in the light of being classified or known about'. These modifications require classifications to be adjusted, corrected and changed, which in turn requires new data and new forms of classification that are in turn self-reproducing. In these ways, Resource Man is not only known, but also made and contested.

Most worryingly, segments are represented as socially factual categories. Industry analysts suggest that 'consumers *sort themselves* into segments with distinct needs and wants' (Valocchi et al. 2007, emphasis added) and that consumers 'belong to segments' (Accenture 2010: 22), rather than recognising that specific demographic and attitudinal questions have been asked by market researchers, based on specific assumptions about who the consumer is and what they value, which lends itself to specific segments and categories. These research findings are then represented as a 'fact-based analysis of consumer behavior, demographics and expressed interest' (Valocchi et al. 2007: 21). In one of very few concessions that consumer expectations may be performed by this research, the IBM Institute for Business Value acknowledges that 'even the numerous consumer surveys focused on consumers' future energy wants and needs, including our own 2007 and 2009 Global Utility Consumer Surveys, may have contributed to expectation setting through questions about a future rich with data, tools for energy usage control, and new products and services' (Valocchi & Juliano 2012: 1). However, the authors still refer to the 'tremendous expectations' of consumers and 'their vision' for the smart grid. The point too often missed is that different ways of knowing consumers leads to the reproduction of different realities (Law 2009). In this case, industry analysts reproduce a reality in which consumers individually manage their energy consumption through an array of ICT- and data-mediated choices.

Empowering Resource Man

In light of the discussion above, it would be inaccurate to suggest that Resource Man is a future utopian fantasy; indeed, in many ways he is already 'real'. Resource Man (and Woman) is increasingly featured on utility websites, where he can be found spruiking the benefits of managing his resources through smart technologies and home energy management systems. For example, the website of the US Pacific Gas and Electric Company (PG&E) features a short video where consumers discuss how they are monitoring their energy usage through the SmartMeter™ data product, which helps them 'pinpoint when they may be consuming more, giving them a better idea of where they can save' (PG&E 2012). There are other signs to indicate that Resource Man is being performed by householders in tandem with the introduction of smart technologies. Leaving aside the ontological and epistemological issues outlined earlier, many consumers say that they want to become a Resource Man, and indeed expect this as an outcome of the smart grid. Over 90 per cent of the 5000 respondents surveyed as part of IBM's 2008 Global Utility Consumer Survey (Valocchi et al. 2009) indicated that they would like a smart meter and associated tools to manage their usage, with 55–60 per cent of those respondents willing to pay a one-time or monthly fee for that capability. Similarly, research conducted by the Smart Grid Consumer Collaborative (SGCC 2012: 16) with US residential energy consumers found that 67 per cent definitely (28 per cent) or probably (39 per cent) would participate in a Smart Meter Data Energy Management programme.

However, other analysts are less enthusiastic. For example, while a Harris poll finds that almost half of Americans (48 per cent) would like to install a 'dashboard' in their home in order to 'proactively manage their energy use', they concede that the likelihood of this actually happening is 'a little soft', particularly given that only 13 per cent say they are very likely to install one (Harris 2012). Similarly, Navigant Research (formerly Pike Research) has downgraded its predictions of Home Energy Management (HEM) users (Pike 2011), describing how the HEM goal of delivering information, visibility and control over energy consumption in the home 'has proven to be elusive', with the promise of the smart home 'largely unfulfilled' (Vyas & Gohn 2012: 1). This report states that many consumers have been 'less enthusiastic about smart meters than utilities originally anticipated', with many programmes failing to move beyond the pilot stage (Vyas & Gohn 2012: 1).

This suggests that a world full of resource men may be further away than the industry or government would like to think. Indeed, for most householders, concepts central to becoming Resource Man are beyond their current understanding. Many do not know what a smart meter or grid is, nor do they understand resource units of kilowatt hours or greenhouse gas emissions; yet these are needed to micro-manage their resource usage (Ipsos MORI 2012; Valocchi & Juliano 2012; Vyas & Gohn 2012: 2; Wimberly 2011; Zpryme 2011). For example, the IBM Institute for Business Value's consumer research notes that 'many consumers around the globe do not understand the basic unit of electricity pricing and other concepts used by energy providers' (IBM 2011: 1). Their 2011 Global Utility Consumer Survey found that over 30 per cent of consumers had not even heard the term 'dollar per kwh' (or the equivalent currency), and over 60 per cent did not know what the terms 'smart meter' or 'smart grid' mean. For more than three in four respondents, the term 'energy portal' had absolutely no meaning (Valocchi & Juliano 2012: 6).

Despite the questionable enthusiasm or ability of householders to become Resource Man, there is now a worldwide endeavour to bring this vision into fruition. Resonating with consumer conceptualisations from the health sector (Halkier & Jensen 2011; Petersen & Lupton 1996), many energy utilities are going about this task by 'empowering' their consumers to get 'energy fit' (SGA 2011: 34) and 'active' (Accenture 2010: 37). This does not necessarily involve going on a resource 'diet', but implies getting smarter and more informed about resource decisions so that appropriate choices can be made. According to Accenture, the critical goal for energy providers is to develop 'a new value proposition that convinces consumers that *extra effort* is worthwhile' (Accenture 2010: 4, emphasis added). Effort is required so that consumers can overcome the 'disconnect between what consumers know about their electricity use, and what they need to know for smarter energy use decisions' (AEMC 2012: 29). Effort is framed in informational and educational terms, where 'high energy literacy' is 'vital' for consumers to navigate the energy market and make 'informed decisions about their tariff options' (Ipsos MORI 2012: 2).

Reiterating the utopian rhetoric discussed in Chapter 2, information is positioned here as 'the fuel that empowers consumers' (Zpryme 2011: 1). Not only will information 'transform the utility industry as we know it today, but it will also transform the way consumers perceive electricity, communicate with their utility, and radically incentivize consumers to become pro-active rather than passive energy consumers' (Zpryme 2011: 1). Evoking prophetic overtones, Zpryme (2011: 1) explains that without information, consumers will be left to make 'random choices' that will hinder 'the rise of the smart energy consumer', who is the 'driving force' behind the Smart Utopia. The humble energy consumer is no longer simply engaged in an economic exchange of paying bills on time; they are positioned as the key to transforming the entire energy industry, and life as we know it.

In assisting consumers to make the necessary effort, utilities, governments and various third parties are positioned as the providers of 'simple and clear information' about energy concepts and terms, pricing tariffs and available choices (Ipsos MORI 2012). This upholds and promotes a linear understanding of change, or a *'consumer energy experience chain'* whereby expectations are driven by perceptions, which are created by knowledge, which 'is retained in the context of core personal *influences* and passed on by trusted influencers' (Valocchi & Juliano 2012: 2; emphasis in original). The crux of this model is that information leads to desired change. Further, change is a fixed goal, something that can be attained and maintained only once the necessary knowledge has adjusted perceptions and expectations. Once again, this model defines energy solely as a resource, commodity or impact.

One of the most popular ways in which energy information is intended to be 'passed on' is through home energy management (HEM) systems and energy portals. Three types of HEM systems are emerging to provide the information and education deemed necessary to elevate Resource Man to his critical status in the Smart Utopia: (i) in-home displays (IHDs); (ii) web-based portals; and (iii) mobile applications, with considerable debate within the industry about which information delivery system is best (Fox & Gohn 2011). Website portal examples include BC Hydro's Compare Your Home and Analyze Your Home tools.[2] IHDs such as the Kill-A-Watt[3] and the innovative Power Hog[4] are sold online and in various electronic and hardware stores, provided by some utilities in association with smart metering programmes, and made available through some local libraries. Online applications include Google's (much-heralded but now defunct) Power Meter and Twitter's tweet-awatt[5] initiative.

Many of these educational efforts are being accelerated through government mandates and industry partnerships. In New Zealand, the provision of an IHD providing simple information on a household's energy usage forms a key component of the country's mandated smart metering deployment for 1.3 million households (PCE 2009). In Europe, the UK and Canada, major utilities have signed deals with companies that will enable and provide HEM systems and services (such as IHDs) to households, and some are offering rebates (or intend to offer them) for customers to purchase IHDs (Fox & Gohn 2011). In the US, nine major energy companies have committed to providing 15 million customers with access to their energy consumption data through the 'Green Button' website portal, which 'can help them reduce waste and shrink bills' (OSTP 2012).

Some programmes that utilise HEM systems provide not only personalised energy consumption information, but also other information services such as energy audits, or other related strategies such as goal-setting. For example, the UK gas and electricity company E.ON's 'Get Energy Fit' programme involves a series of actions and commitments similar to an exercise or fitness programme. Described as a 'simple but very effective education campaign' (SGA 2011: 34), it involves a three-step process of taking a survey, starting an energy dashboard, and setting 12-month goals. The programme also encourages community comparisons, which are noted as 'powerful catalysts for change' (SGA 2011: 34).

This last aspect of the Get Energy Fit programme encapsulates a very loose recognition that *homo facultas* (Resource Man) is also related to *homo sociologicus* (Social Man). Similarly, consumer reports recommend that utilities 'tap into people's inherent social nature' (IBM 2011: 1) through portals that allow consumers to see and compare their usage with that of others. This acts as a 'social action trigger' that enables consumers 'to determine the right ways to act in many situations' (IBM 2011: 1). The idea here is that utilities can 'plug into' the social compartment of Resource Man's brain, much as an appliance is plugged into an electrical socket. Of course, some consumers *do* compare and benchmark their energy usage and this has led to real energy reductions. For example, the energy data management company Opower[6] has had good success with these strategies using their Opower Energy Social Application. However, many people have no interest in their energy data and very little understanding of what they might do with it or why they would want to tell their friends about it (Strengers 2011b, see Chapter 5). These 'social triggers' encourage those who are interested to compare themselves – as Resource Man to Resource Man, not as people who use energy to do laundry, cooking, shopping, bathing or any other activity (although these issues do slip into discussion forums and workshops). In this way, 'social' understandings of the consumer maintain a commitment to the Resource Man vision, encouraging consumers to 'tweet' their kilowatt hours to their friends, or sign up to benchmarking programmes where people share their experiences *with* and *about energy* (as a resource, commodity or impact). This represents an incredibly narrow perspective on human experience.

Smart one word

In contrast to empowering Resource Man to become an efficient manager of his resources, proponents of the Smart Utopia also envision a new smart life for this consumer, whereby smart technologies and tools help to maintain or improve his lifestyle, with minimal or no effort. In this passive aspect of the consumer vision, Resource Man assigns resource management and a range of other household practices to smart technologies, which do energy management 'work' and domestic labour on his behalf. A range of technologies fall into this category, such as smart appliances, home automation systems, programmable thermostats, and direct load control (see Chapter 7). Proponents of micro-generation and electric vehicles also emphasise the idea that Resource Man can decarbonise his lifestyle through technological substitution (see Chapter 8). By producing his own renewable power, driving an efficient or renewable-powered electric vehicle, automating appliances or assigning control of an appliance to a utility to turn on and off on his behalf, Resource Man is able to continue doing what he does, just in a 'smarter' way. He is also able to circumvent much of the 'effort' and 'action' required by the methods discussed above.

The attainment of a smart lifestyle is problematic for a number of reasons, not least of which is that it encourages and enables new forms of consumerism, with its associated resource usage and e-waste

(Pamlin 2002). Reflecting international trends, the consumer electronics sector is now the single-most significant growth area of UK domestic electricity consumption, and by 2020 it is predicted to be the biggest single user of domestic electricity (45 per cent of household energy) (EST 2007: 3). The Energy Savings Trust (EST 2007) notes that 'higher spec' versions of electric gadgets have tended to consume more energy than they replace, fuelling further electricity demand. Smart appliances and HEM systems are part of the broader domain of consumer electronics that provides 'incremental improvements' and 'seemingly reinvent[s]' old ideas in ways that can in fact increase energy demand (EST 2007: 8).

Attempts to sell new smart products and services to Resource Man are coming from a diverse group of interests that extend beyond energy providers and governments, such as energy service companies (ESCOs), appliance manufacturers, ICT companies (including HEM and home automation businesses), and housing developers. Tellingly, some of these companies feature in the electricity industry's history as the purveyors and promoters of past energy visions. General Electric (GE), for example, previously led the charge to foster 'a new electrical consciousness' in America, expressed as 'the desire of individual families to make their homes into electrified dwelling places' (GE in Healy & MacGill 2012: 34). In the Smart Utopia, GE has reinvented itself as a key innovator and promoter of smart home appliances. Encompassing the broader techno-optimism central to technological utopian visions, GE promises that technological innovation can 'meet today's environmental challenges while driving economic growth'.[7] The ideals of 'smart living' promoted by companies such as GE aim to increase the efficiency of achieving 'improved' entertainment, information or comfort. In other words, Resource Man might be managing his standby power on a home entertainment theatre, rather than a small analogue television.

This makes the Smart Utopia's aim of shifting and shedding demand 'whilst having minimal impact on day to day habits' (Jelly 2008: 66), 'in a way that avoids significant impacts on comfort and lifestyle' (Reidy 2006: iv), incredibly dubious, or at least highly problematic. Indeed, the ultimate aim and agenda for many smart utopians is achieving 'lifestyle improvements', not merely avoiding impacts on lifestyle (SGA 2011: 20). Resource Man might have unprecedented access to ICT tools that manage his consumption and empower him to make decisions about his consumption; but he is also implicated in realising a new smart lifestyle in which electricity is even more essential for achieving climate control, enhanced security, unparalleled access to entertainment, and heightened levels of relaxation. Resonating with Shove's (2003; 2004) analyses of technologies and consumption, the risk here is that a fully automated, climate controlled, efficient smart home can ratchet up energy demand, so that householders use more rather than less, under the guise of 'smartness'.

Engaging with Resource Man

Another way in which history is re-enacted in the smart energy vision is through the sorts of energy relationships envisaged for Resource Man. As already alluded to, this consumer conceptualisation positions energy consumers as energy apprentices who require expert energy knowledge and skills from an energy 'master'. This expert–learner relationship maintains the 'paternalistic culture' of utility industries (Tom Standish, Chief Operating Officer of Centrepoint Energy in Honebein *et al.* 2009: 40), and narrows the types of partnerships and collaborative relationships possible in the Smart Utopia. It also reflects a broader commitment to the traditional provider–consumer relationship, which has 'evolved at a glacial pace' (Accenture 2010: 4). A similar point is made by Marvin *et al.* (2011: 185), who find that an 'authoritative relationship' is often employed in UK smart metering programmes, where utilities seek applications that allow them to exert 'centralized control over their customers' consumption'. Even where new relationship models are emerging which position energy consumers as 'producers', 'prosumers' or 'co-managers' of their energy demand and supply, technology and data still mediate and manage this relationship.

Nonetheless, there is a growing recognition from both providers and consumers that the rules of engagement must change to achieve the Smart Utopia's aims. For example, US consumer research suggests that some consumers 'are looking for alternatives to traditional service' and 'welcome higher levels of engagement with their electricity suppliers', with smart grids cited as the 'enabling platform' (Wimberly 2011: 3, 21). Similarly, an IBM consumer survey conducted globally found that almost 70 per cent of 5000 respondents were willing 'to take advantage of what might be offered in a partnership that differs from the traditional utility-customer relationship' (Valocchi *et al.* 2009: 8). Some analysts argue that changing provider–consumer roles will be an inevitable outcome of the smart grid, that 'this new system will ... transform the relationship between the utility and consumer from a one-way transaction into a collaborative relationship that benefits both, as well as the environment' (Peter Corsell, CEO of GridPoint quoted in WEF 2009: 3).

Other analysts cite the historically passive role of the consumer as a major impediment to change:

> 'It sounds simple and trivial today – but the idea of getting consumers to become active participants in the market is still novel to many in the industry and even more so to the average consumer who has been successfully trained to be a passive user'.
> (Sioshansi 2012: xxxvi)

This goes both ways. Providers have traditionally been just as uninterested in consumers as consumers have been in providers, with the realm past the meter being 'figuratively and literally beyond [the utility's] ... control, influence, or interest' (Sioshansi 2012: xxxiii). However, in the Smart Utopia, the consumer becomes 'an important part of the resource mix' – an innovator and service provider in their own right through their uptake of embedded generation, demand response and energy conservation (Wimberly 2011: 21). In other words, they are intended to become part of the energy supply chain, using the same energy management tools as utility providers.

While most relationship models envisage consumers at the end of this supply chain, progressive industry research on the evolving provider-consumer relationship suggests that consumers will move further up the supply chain, 'becoming designers, producers, marketers and distributors of the products they once just purchased' (Valocchi *et al.* 2007: 3). Similarly, others suggest that consumers will become 'codesigners' or 'co-creators of value', contributing by having 'a role in generation requirements, designs, scripts, and prototypes for a product or service' (Honebein *et al.* 2009: 40). Recognising that a number of third parties are now intervening in the provider–consumer relationship, Accenture suggests that utilities may need a 'partnership ecosystem' that includes 'engagement with cities, municipalities, governments and other stakeholders to drive the development of projects such as intelligent cities' (Accenture 2011: 42).

While this sounds promising, smart technologies such as direct load control (the remote control of appliances) are also a method of potentially strengthening the paternalistic relationship between providers and consumers of power, whereby utilities seek and gain new methods of discipline and control (Marvin et al. 2011). Concerns over who or what is control, as well as how data is being collected, stored and for what purpose, have led to suggestions that the smart meter is a 'Big Brother' technology (Vermeer 2008) or a 'spy of the home' (Daily Telegraph headline cited in Marres 2012b: 300). Similarly, in US consumer research, almost half of the respondents identified the electric utility's capability to monitor and control electricity usage as a reason for their unfavourable opinion of smart meters (Vyas & Gohn 2012). This resonates with Akrich's (1992: 317) observation that 'the set of meters is a powerful instrument of control',[8] which reflects the current relationship between providers and consumers of electricity.

It is also important to remember that even in collaborative or expanded relationship scenarios, the provider maintains their role as the expert and the educator, with consumers' participation largely dependent on them gaining energy management knowledge and skills. The language of an IBM report is revealing in this regard, instructing energy providers to 'leverage consumers' newfound openness to change, and then 'provide information, influence behavior and teach consumers new ways to meet their goals' (Valocchi et al. 2009: 4). Similarly, in recognising the need to 'restore the relationship' between the energy industry and customers, the CEO of a smart technology supplier recommends that 'the utility of the future needs to demonstrate that they can *help customers manage their energy better*' (England 2012: 1, emphasis added). The vision of Resource Man is again implicit in these remarks, where utilities are framed as the expert providers of energy-management technologies and information. Where consumers contribute 'their knowledge, skills and attitudes', they do so within the scope of expert resource management understandings of energy products and services to meet their 'rational' and 'emotional' needs (Honebein et al. 2009: 40–1). The point here is simple: where consumer reports refer to new energy relationships and modes of engagement with people who consume energy, they do so with Resource Man firmly in mind.

In summary, while there is broad industry agreement that changing the relationship between providers and consumers will be a necessary step in achieving Smart Utopia, the relationship models put forward by industry continue to perform a paternalistic relationship mediated by and through technology and data. As Marvin *et al.* (1999) suggest, it is likely that a number of different relationship models will emerge, overlap and be contested as the smart energy vision takes shape. However, what currently binds them together, at least in the reports analysed here, is the belief that it is Resource Man who the energy industry will and should be engaging with in the future. This precludes not only alternative understandings of social action and change, but also alternative ways of imagining provider–consumer relationships.

Setting Resource Man aside

This chapter has demonstrated how research about and with the new energy consumer performs a specific and homogeneous conceptualisation of people and their future role in the Smart Utopia. The ultimate energy consumer, with whom energy utilities are seeking to engage in new programmes, partnerships and relationships, is Resource Man – a data-driven, information-hungry, technology-savvy home energy manager, who is interested in and capable of making efficient and rational resource management decisions. I have argued that this productive vision of the consumer represents a narrow vision of human action and experience, and promotes a new energy-intensive smart lifestyle that may undermine the Smart Utopia's aims of decarbonisation and reduced peak demand. If we are to imagine different possible realities for consumers, including different ways in which people relate to energy, technology, utilities, and indeed the world, we need an alternative ontology (or ontologies) of social action and change.

In the following chapter, I put aside the current fascination with Resource Man to depict a different ontology of consumption and everyday life. Drawing on theories of social practice, I frame energy as a material participant in the practices performed in the home, such as heating, cooling, laundering, entertaining, cleaning and so on. Rather than being a unit of resource consumption, a commodity to be bought, or an impact to be managed, I reposition energy as the stuff which makes many domestic practices possible, and which in turn can reproduce and reconfigure them. The question then becomes not (only) about how smart energy technologies will be adopted or accepted by consumers, how they will change their individual consumption habits and decisions, or even how they will be domesticated into the home. Instead, I ask how energy and smart energy technologies are integrated into everyday practices, and what this means for the ways in which social action and change is constituted.

4
Energy in Everyday Practice

It is time to put the smart ontology to one side, and introduce another possible reality (or series of realities) for smart energy technologies in everyday life. This chapter depicts an ontology of everyday practice in which people who consume energy are repositioned as performers of everyday practices, who are in turn enrolled in realising, or undermining, the aims of the Smart Utopia. This is an ontology in which social order and change are grounded in the routines and dynamics of day-to-day living, where people consume energy by undertaking everyday practices such as cooking, cleaning, bathing, cooling and entertaining. In developing this position I pay particular attention to the ontological status of energy and the infrastructures and technologies that make and mediate it. I am interested in how energy, along with the technologies that deliver it to the home and that it in turn powers, can be thought of as material elements of practice, or part of a practice's composition. I begin this discussion with a brief and final departure from the smart ontology by outlining the ways in which everyday life is commonly dismissed from the Smart Utopia, where it is inadvertently and implicitly framed as dumb and disorderly behaviour.

Dumb and disorderly behaviour

One of the critical gaps in the smart ontology is its oversight of the dynamics of everyday practice. Anything that is not characterised as 'smart', namely any activity that is not mediated by information and technology in a rational and instrumental manner is, by implication, positioned as 'dumb'. Consequently, there is an absence of the 'messiness of everyday life' (Dourish & Bell 2011: 4), of the daily domestic routines involved in preparing meals, cleaning the body, clothes and homes, or making spaces and people comfortable. These routines are of course features of the Smart Utopia, but their achievement is a matter of technological mediation, substitution and information management. By inadvertently positioning everyday practices as 'dumb' or 'messy', the smart ontology reduces a significant proportion of human experience to one or both of these disorderly states, in which order is either absent through a lack of intelligibility (dumb) or through the absence of any ordered or discernable patterns (mess). Both concepts imply chaos and disharmony; an apparent lack of order in an otherwise ordered reality.

Dumb and disorderly behaviour confusingly and disturbingly characterise a raft of social, cultural and behavioural 'issues' or 'range of factors' in the Smart Utopia which are commonly compiled together into large lists (AEMC 2012: 24). For example, one report cites such factors as 'existing habits, social norms, behaviours and attitudes...the ability to process information, price of products and services, awareness of energy costs, availability of time, access to finances, and general appetite [for] or commitment to change' (AEMC 2012: 24). Lists of these factors are potentially inexhaustible, and frame human diversity as an issue, or as a series of problems or 'barriers' that need to be addressed and eliminated through ICT and data-related tools and techniques.[1] This dumb and disorderly behaviour is presented as a challenge for energy utilities and policymakers – a constant spanner in the work of their unrelenting endeavour to achieve reliability and efficiency – and something that must be 'taken into account'. In this way, any activity that is not characterised as 'smart' is either ignored or 'solved' through data and ICT solutions.

As I argued in Chapter 2, clearing away or attempting to remedy non-smart activity is a deliberate and arguably necessary move common to modernity projects, whereby social complexity is managed by reducing it to manageable and actionable chunks or segments, with targetable groups of people and clear tools and techniques to engage them. Nonetheless, this move generates two significant problems; first, it reduces and often dismisses a significant proportion of human (and non-human) activity by positioning it as a series of unrelated and unordered factors; and second, it classifies such activity as something that both providers and consumers of energy should ignore, seek to eradicate or 'solve' by transforming 'dumb' activity into 'smart' activity.

This is a highly problematic position. For example, Dourish and Bell (2011) warn that the ubiquitous computing vision has so far failed to avoid or eradicate the mess of everyday life, despite its best intentions. Instead, they call for researchers to embrace and value messiness as 'inspiring, productive, generative, and engaging' (Dourish & Bell 2011: 93). They argue that mess is not something to be fixed, tamed or removed; indeed this is an impossible goal. Rather, messiness is 'dynamic, adaptive, fluid, and open' – it is the stuff upon which innovation, improvisation and adaptation can be founded (Dourish & Bell 2011: 93). The challenge is not to eradicate mess, but to account for, understand, embrace and conceptualise it as more than a 'range of factors'.

The remainder of this chapter moves past the clear-cut binaries of disorder and order, smart and dumb, which are implicated in the smart ontology. Instead, I draw on theories of social practice to understand the order and continual reorder of everyday activity, and the role of energy and its associated technologies in these processes. My attention remains focused on the practices of the home, where the Smart Utopia is intended to be realised.

Introducing social practices

'Practice' or 'practices' are overused terms which encompass virtually every understanding of human (and non-human) action. Definitions vary widely between disciplines. Many position 'practice' as unwaveringly human, where technology is either completely absent, or variously positioned as a symbolic conveyer of meaning, a status symbol, or an agent of change. 'Practice' is also often used synonymously with 'behaviour', being seen as the action arising from the attitudes, values and opinions of autonomous individuals, or the outcome of social and cultural norms to which individuals subscribe. This is partly because behavioural theories remain the single-most dominant method of understanding social action and change (Shove 2010a). These theories also underpin the smart ontology, where rational choice theory and information-deficit models begin from the position that individuals require information, price signals and other tools to make informed decisions about their energy consumption (see Chapters 2 and 3). Given the dominance of behavioural theories, there is a tendency, and risk, for all other theories of social action, including theories of practice, to be collapsed or interpreted within them.

In contrast, I draw my definition of 'practice' from theories of social practice, which are rooted in the disciplines of philosophy, sociology and anthropology (Bourdieu 1977; Giddens 1984; Reckwitz 2002b; Schatzki 2002; Shove *et al.* 2012), and date back to the philosophers Wittgenstein and Heidegger (Shove *et al.* 2012). While practice theories have an eclectic history spanning multiple decades or two 'generations' (Postill 2010), they are united in the assertion that practices 'ordered across space and time' (Giddens 1984: 2) are the foundation of social order. In other words, practices, rather than people, data or technology, are 'the whole of human action' (Reckwitz 2002b: 249). This is a fundamental distinction between the smart ontology and an ontology of everyday practice.

Delving further, theories of social practice collapse common binaries between individuality and social totality, and agency and structure. People are positioned neither as autonomous rational actors nor as social dupes (Reckwitz 2002b). Similarly, structure is neither a force that acts upon society nor a product of social forces. Rather, practices constitute what Giddens (1984: 24) describes as a 'duality of structure', whereby 'the day-to-day activity of social actors draws upon and reproduces structural features of wider social systems'. In the context of the Smart Utopia, this means that when people participate in everyday practices that consume energy, such as making a cup of tea, taking a shower, or using the air-conditioner, they simultaneously draw on and reproduce the structural features of an energy system. In this sense, the properties of energy systems 'are both medium and outcome of the practices they recursively organize' (Giddens 1984: 25).

A related point is that practice theorists understand innovation and change as occurring through and within practice, resulting in practice 'formation, reproduction and dissolution' (Pantzar & Shove 2010b: 450). Thus the Smart Utopia (which is itself the product of a suite of practices) is imagined, unfolding, and taking or not taking shape through its integration into (and emergence from) the practices that it supports and performs.

Many practice theorists place particular emphasis on the human body as the conduit or site of the performance of practices or routines, which are 'the product of training the body in a certain way' (Reckwitz 2002b: 251). Similarly, Schatzki (2001: 2, emphasis added) suggests that practices are *'embodied*, materially mediated arrays of human activity centrally organized around shared practical understanding'. Resonating strongly with Bourdieu's notion of *habitus* (Bourdieu 2005), the body is positioned as the intersection between agency and structure; it is the carrier of histories, sociotechnical structures and a sense of place (Hillier & Rooksby 2005). It is also through the body that people, as carriers, performers or practitioners (Pantzar & Shove 2010b; Reckwitz 2002b; Warde 2005), perform practices, and it is through performance that practices exist in the world and are reproduced and transformed (Shove *et al.* 2012). It also through the enactment of practices that different ontological realities are performed in everyday life (Law 2009). In this way, practices are both performances and performative.

As well as being understood as a bodily performance, practices are also understood as mutually intersecting entities (Shove *et al.* 2012). A practice entity is what we readily identify as a practice (activities such as cooking, swimming or cycling). A practice entity may be durable, with a path and trajectory of its own, but it is by no means fixed or static (Warde 2005). Rather, practice entities are reproduced and transformed through their performance in everyday life. A practice entity can be conceptualised as being composed of several overlapping elements; theorists differ in their understandings of these (Schatzki 2001). In this book I follow Shove *et al.*'s (2012) simple model, which proposes that a practice comprises three intersecting elements: skills, meanings and materials.

The first element quite literally refers to the competence or skill needed to perform a particular practice. Other practice theorists and researchers use related terms to mean similar things, such as 'practical understanding' (Schatzki 1997), 'practical intelligibility' (Schatzki 2002), 'procedures' (Warde 2005), 'know-how' (Reckwitz 2002a), 'practical knowledge' (Reckwitz 2002a) and 'knowing in practice' (Wenger 1998). The key idea here is that knowledge is embodied and acquired through lived experience, or through practice. The second element is the meanings, images (Shove & Pantzar 2005a) or understandings (Warde 2005), which encompass ideas about what is right, proper, normal or acceptable, such as when it is acceptable to wear a stain on a piece of clothing and when it is not. The third element, materials, is variously described as 'artifacts, hybrids and natural objects' (Schatzki 2001: 2), 'artefacts or things' (Reckwitz 2002a: 208), 'requisite material arrays' (Shove & Pantzar 2005b: 59), 'material infrastructures' (Strengers & Maller 2011) or more simply the 'stuff' of everyday practice (Shove *et al.* 2007).

Dividing practices into a set of elements as described above is an 'oversimplification and an abstraction', but one that is often considered useful for analysing and understanding the composition and dynamics of specific practices (Pantzar & Shove 2010b: 453). This simple model suits my purpose, which is to understand how smart energy strategies are shaping, reshaping and disrupting everyday practice. In summary, I take a practice to be a constellation of elements – materials, meanings and skills – that are linked together to form a recognisable entity (cooking, showering, laundering) that is performed and transformed *'through* the process of doing' (Shove *et al.* 2012: 41, emphasis in orginal).

One other clarification to make here is that practices do not always exist in isolation, but can congregate together in loose-knit 'bundles' or 'stickier' and more tightly linked 'complexes' (Shove *et al.* 2012: 17). This is particularly important in relation to everyday practices that consume energy, many of which are woven together in various temporal routines within the home. For example, cooling or heating the home is often linked to a raft of other practices, such as cooking meals, working from home, entertaining guests, or caring for sick or young people (Strengers & Maller 2011). Similarly, the Smart Utopia enables the emergence and integration of new materials, meanings and skills into a variety of everyday practices, which may transform entire bundles and complexes of practices rather than single and isolated entities (Shove *et al.* 2012).

As introduced in Chapter 1, my use of the term 'everyday practice' signifies my focus on the limited suite of practices performed regularly and routinely in order to carry out daily activities. As such, I follow modern scholars' broad use of the 'everyday' category, particularly Pink (2004, 2012b), Shove *et al.* (2007) and Michael (2006). While these authors differ in their specific orientations to the everyday, they are united in positioning this domain as highly dynamic, with technological artefacts implicated in 'everyday processes of ordering, reordering or disordering' (Michael 2006: 38). These authors simultaneously position the everyday as mundane and ordinary as well as vibrant and potentially transformative. My use of the term resonates with Michael's desire to move beyond futuristic depictions of technology at the expense of their already performative role in everyday life. I take Shove and colleagues' (2012; 2007) theorisation of practice as the basis for thinking through how ordinary objects both sustain and transform the dynamics of the everyday, and I situate my analysis by borrowing from Pink's (2004, 2012b) use of the term, where the everyday refers to the site where life is lived out through engagement in a series of domestic practices. My analysis is further limited by my focus on everyday practices in the home that consume energy: it is here that we find important theoretical gaps in understanding how energy, and the technologies and infrastructures it relies on for production, distribution, transmission and use, fit within an ontology of everyday practice.

Practice theory has had little to say about the role of intangible or 'immaterial materials' (Pierce & Paulos 2010) such as energy, instead viewing energy or consumption more broadly as an *outcome* of practice or 'a moment in every practice' (Warde 2005: 137). This position is based on the premise that 'people do not consume energy *per se*, but rather the things energy makes possible, such as light, clean clothes, travel, refrigeration and so on' (Wilhite 2005: 2). Following this position, a number of researchers, including myself, have studied the emergence, spread and transformation of practices-that-use-energy (Gram-Hanssen 2009, 2010; Hitchings 2007; Maller 2011; Maller *et al.* 2011; Røpke *et al.* 2010; Strengers & Maller 2011). With the focus shifted from energy *per se* to the practices that use it, non-'expert' forms of knowledge take precedence, as do the materialities, meanings and routinisation of lived experience in which energy participates. There have been many worthy attempts at explaining and conceptualising the ways in which energy is experienced or known through and in practice, and it is to these ideas that I now turn our attention.

Making sense of energy

There is significant ambiguity about what energy actually is. In the energy industry, energy is defined in relation to its production and consumption, where it features as a tradable commodity and a producible and measurable resource with a series of definable and quantifiable impacts. Household energy bills also define energy in these terms, where it features as a commodity (price), resource (kilowatt hour) or impact (greenhouse gas emissions). However, these are not the only ways in which energy is known.

In everyday life, energy has a plurality of meanings which overlap with and extend the energy industry's framing of it. Energy is variously featured as a tradable commodity, a precious resource, a social necessity, an indicator of social progress, or in physics, an indirectly observed quality (Pierce & Paulos 2010). Energy is also commonly referred to in relation to the body, where it is used to describe different degrees of emotional, physical and mental energy. Energy's complicated history leads Pierce and Paulos (2010) to remark on its 'ambiguous ontological status' and its seeming intangibility in everyday life.

Tackling this issue, sociological and anthropological research on energy consumption has continually sought to understand how people make sense of energy in their everyday lives (Shove *et al*. 1998; Wilhite *et al*. 2000). To take one example, the media anthropologist Pink (2012a: 121) finds that people understand energy through the 'affective and multisensory feelings' reproduced in domestic energy-consumption practices. Using methods such as walking video tours of the home, autoethnography (energy use diaries) and following domestic artefacts (such as laundry) around the home, Pink develops a 'sensory ethnography' which focuses on the qualities of the way energy is experienced through the senses (smell, taste, sight and sound). Pink's account allows for a nuanced understanding of how a largely invisible resource like energy moves through and around the home, and how it is embodied in places and things.

Other researchers have argued that people have different 'folk understandings' (Kempton & Montgomery 1982), 'energy sensibilities' (Berker 2013) or 'domestic multi-cultures' (Sofoulis 2011) that help them make sense of energy in their everyday lives. For example, in a study analysing user decisions about when to turn on an air-conditioning unit, Kempton *et al*. (1992a: 189) found that their research participants drew on a range of 'folk physiological theories' about their body types and needs, which resulted in three-quarters of the sample bypassing built-in thermostats, thereby thwarting the

Energy in Everyday Practice 61

Resource Man aspirations of utility providers. Similarly, Wilk and Wilhite (1985) conducted research to understand why householders don't weatherise[2] their homes when there are such clear economic paybacks. They found that every homeowner interviewed was aware of the costs of heat leakage as well as the benefits of weatherisation, despite not doing it. They explained this 'irrational' lack of interest in weatherisation through folk theory which emphasises the health benefits of fresh air and weatherisation's lack of glamour and visibility when compared with other environmental actions, such as installing solar photovoltaic panels.

These links to health knowledges remind us that expert and folk understandings about energy exist in tandem, and further, that they are not always about energy. For example, Petersen and Lupton (1996: 51), citing Davison *et al.* (1992: 678) discuss how 'lay epidemiology', which is based on 'the routine observation of cases of illness and death in personal networks and the public arena', can compete with expert epidemiological knowledge about the causes and cures of disease. This lay and expert knowledge also manifests itself in the everyday practices of the home, where understandings of and recommendations for creating a 'healthy', 'comfortable' and 'safe' environment are integrated into cleaning, bathing, laundering, heating, cooling, cooking and entertaining practices, with significant implications for energy demand (Strengers & Maller 2011). Incidentally, there is also a growing resistance to smart meters founded on health concerns.

Similarly, in theories of social practice, the focus is not on energy itself, but rather on the elements of the practices-that-use-energy, such as the meanings of what it takes to produce clean homes, skills related to how to prepare a meal, or knowledge of the materials required to cool a house. Energy has a much more ambiguous role in these dynamics of practice, as do the meanings and skills needed to know and handle it. In order to develop this conceptual territory, the discussion below builds on the position that energy is a material of social practice, with distinct qualities that intersect with other practice meanings and competencies to inform 'what makes sense to them [people] to do' (Schatzki 2002: 75). This is another simplifying move that allows us to understand how matter – in this case energy and its associated technologies – comes to matter in everyday life (Barad 2003). More specifically, it allows us to think about how smart energy technologies and the energies they produce participate in and reproduce everyday practice.

Energy in practice

The concept of energy that I develop below builds on a post-humanist strand of social practice theory emerging primarily from science and technology studies (STS). Pickering (1993, 1995) was one of the earliest theorists to put forward a materially oriented theory of social practice as an alternative to actor-network theory (which is also proposed as a way of decentering the human subject and of thinking semiotically or symmetrically about human and non-human agents (Latour 1987a; Law 1993)). Pickering (1993: 567) proposed that 'the trajectories or emergence of human and material agency are constitutively enmeshed in practice by means of a dialectic of resistance and accommodation' – a process he describes as 'the mangle'. Pickering's emphasis on 'mutually and emergently productive' human and material agencies challenges the symmetry between human and material realms proposed by semiotics, and goes a considerable way to demonstrating the dynamics or the continual 'tuning' of materials and humans in practice (Pickering 1993: 567; 1995).

Making further sense of this mangle has been of concern to modern social practice theorists, who have sought to extend Latour's and other STS scholars' ideas that social action is made up of human and non-human actants (Latour 1987b). Schatzki (2010: 129) accounts for materiality in his ontology of practice as 'material arrangements' or, rather, a 'set of interconnected material entities' ('humans, artifacts, organisms, and things of nature') that are connected to practices, but nonetheless distinct from them. Shove *et al.* (2012; 2007) go further, arguing that materials are an element of practice – that is, they are *a part of practice itself*. These scholars suggest that even when materials appear stable or 'fixed', 'their social significance and their relational role in practice is always on the move' (Shove *et al.* 2007: 8). This position extends understandings of materiality which posit artefacts as constructing socialness, things as carriers of social or cultural meaning, or objects as more or less stabilised entities once 'appropriated' or 'domesticated' into practice (Shove *et al.* 2007). Instead, this conceptualisation allows for dispersed forms of agency to be recognised, 'where the

Energy in Everyday Practice 63

more-than-human or hybrid dynamics of meaning and matter are central' (Hawkins & Race 2011: 114).

Reckwitz (2002a: 208) adopts a similar position, suggesting that materials and technologies 'necessarily participate in social practices just as human beings do'. Building on this contention, Reckwitz (2002a: 212) suggests that 'certain things act, so to speak, as "resources" which enable and constrain the specificity of a practice'. While Reckwitz (2002a) is not referring to 'resources' here in the sense of energy or water, I would argue that it is possible to conceptualise them in this regard: that is, as materials that participate in practice. This definition is distinct from the term's use in the smart ontology, where 'resources' refers to the rational and technologically-mediated management of natural resources in isolation from, or separate from, their integration into the practices they enable. In contrast, Reckwitz's (2002a) concept of resources and Shove et al.'s (2007) account of materiality point towards the role of energy and its associated technologies as being within and integral to everyday practice.

These accounts take us a significant way towards an understanding of materiality in practice; however, they do not go so far as to inscribe energy with material status, or to articulate a clear role for it in practice. Indeed, the relationship between energy, its technologies and practice has yet to be worked through in significant detail. Nonetheless, we have the theoretical resources to do so. Hawkins and Race (2011: 116) provide inspiration in their research on bottled water, where they suggest that a focus on bottled water practices 'pays close attention to the ontological realities of bottles in action'. Similarly, we can think about how the everyday practices-that-use-energy requires us to pay close attention to energy (and its associated smart energy technologies) *in action* and, conversely, in *inaction*. Reiterating a point made above, I am not primarily referring to energy's 'action' in terms of its use as a commodity or natural resource, but to its ability to 'enable and constrain the specificity of a practice' (Reckwitz 2002a: 212). I am interested in how energy, as a 'resource' that is integrated into practice through a variety of mediating technologies (such as appliances, meters or smart control devices), enables and constrains what it means to do laundry or cool the home, for example, or otherwise makes 'demands' on, or has demands placed on it by, other elements of practice.

Building on the ontological reality of energy in action, there are grounds to consider energy, or different *energies*, a material element of practice (Strengers & Maller 2012). This position is distinct from other STS accounts of practice which have focused their attention on specific and tangible material things, such as cameras in the practice of photography (Shove & Pantzar 2007), new media or ICT technologies in media practices (Røpke *et al*. 2010), and patio heaters in outdoor comfort experiences (Hitchings 2007). With the focus on these material things, larger systems of resource provision and their associated technologies have had a more ambiguous status in practice, being variously positioned as 'intermediaries' between the providers and consumers of resource systems (Marvin *et al*. 1999; Southerton *et al*. 2004; Van Vliet *et al*. 2005), the 'connective tissue' that binds providers and consumers into 'distinctive regimes of resource management' (Chappells & Shove 2004b: 142), or in the case of appliances and power sockets, the 'terminals' or 'sensitive fingertips of existing infrastructures' (Shove & Chappells 2001: 57). While these accounts provide valuable and commonly unacknowledged links between production and consumption, they do not position the energies produced and delivered by large or small-scale energy systems, smart grids or other energy technologies as part of practice itself.

Identifying this gap, Maller and I (Strengers & Maller 2012) have developed an account of materiality in which we argue that different 'energy-making practices can create distinct object-like energies... [that] intersect with other elements to reproduce (or limit) resourcefulness'. The term 'energy-making practices' refers to the practices of making usable energies, such as chopping wood for a fire or using a micro-generation unit. For example, in some cultures, making energy (and water) involves distinct arrangements of 'technical "equipment" (donkeys, axes, used to collect and store these resources), practical knowledge, and understandings' that render these resources usable for practices such as cooking or heating (Strengers & Maller 2012). These in turn give different energies specific meanings and values: a precious resource that should not be wasted, or which should only be used for certain purposes, for instance. We conclude that the qualities of different energy and water systems define how energies (and waters) are constituted and integrated into practice as a material 'thing' or 'things', and therefore the demands and dependencies they make and create in practice. A related implication is that the service relationships between providers and consumers are implicated in the types of energies being produced, which in turn have repercussions for how energy is consumed. Pierce and Paulos (2010: 117) make a similar point, arguing that differentiations and distinctions between the types of energies that are prioritised and manifested through different energy technologies and strategies are important to ensure that energy does not only enter our everyday experience 'as a single, totalizing entity or phenomena – something vague and amorphous with which our only real concern is "connecting to"'. Noting that energy is commonly referred to in the singular 'by design', they view design as the path to reintroducing its plural form – energies – by manifesting and (re) designing different types of energies (Pierce & Paulos 2010: 117). This involves paying attention to how energy is 'done' (Lien & Law 2011), how it is 'made' and how it is made meaningful as the material thing or things we know and experience as energy through the practices of everyday life.

While these are promising lines of enquiry, thinking about energy as a material of practice remains a tricky business, partly because energy has an often intangible material or physical presence in practice, often noticeable only through its absence or invisibility. Schatzki (2010: 125) offers a way out of this bind, suggesting that materiality does not only connote physicality, but refers to the 'stuff' of social practice, or its composition. He develops this understanding in relation to biophysicality and the environment; however, it could equally extend to other elusive materialities of practice, such as energy. Bringing this together with Shove *et al.*'s (2012; 2007) conceptualisation of materiality, we can understand energy as part of the composition of practice – a material element that sometimes makes demands on practice through its immateriality, or through other tangible materials, such as appliances, light globes or solar panels.

From this perspective, the technologies and strategies of the Smart Utopia are absolutely essential in understanding what energy is and how it comes to matter (or not matter) in everyday practice. We might then ask: How does energy feedback and dynamic pricing change the meanings of energy in practice? How are home automation technologies integrated into everyday routines? And how do different energy-making practices, such as those involved in micro-generation, produce different energies that come to matter in everyday practice? These are some of the lines of enquiry I pursue in Part II of this book.

However, there are still questions that remain here, not least of which concerns energy's hazy ontological status in relation to the technologies and infrastructures that produce it, deliver it to homes, and mediate its consumption. For example, does energy-as-material have the same status in practice as an air-conditioner, hairdryer or toaster? Probably not. Does focusing on energy-as-material simply perpetuate the dominant paradigm in which energy, rather than materials, meanings or skills, becomes the dominant focus and point of potential intervention in facilitating change? This is a definite risk. However, energy *does* appear to play a material role in practice, even if this role is still theoretically and conceptually unclear.

The point of elevating energy to the status of a material, however elusive, is this: it allows us to analyse how the changing materialities of energy that emerge through smart energy strategies order and reorder everyday practice. If we think of energy as an element of practice we can examine what this material means (in practice), how it is positioned in relation to other elements of practice, how its role in practice is changing – or has already changed – through the introduction of smart energy technologies, and how energy-as-material is implicated in transforming practice. Of course, this does not mean that we should lose sight of the other materials of practice, many of which feature in the subsequent pages. The materialities of in-home displays, smart appliances and programmable thermostats also deserve our attention, and are given it, in the analysis that follows in Part II of this book.

A note on energy practices and consumers

Before continuing, it is important to clarify my avoidance of the term 'energy practices' in the chapters that follow. Despite at least half this chapter being devoted to the subject of energy, I wish to de-emphasise the role of energy (as a commodity, resource of impact) in practice, or rather I do not wish to define these practices in relation to energy. Instead, I use the terms 'everyday practices', 'household practices' or 'practices-that-use-energy'. Laundering and showering, for example, which are both practices-that-use-energy, could equally be performed without energy, albeit somewhat differently. Thus to identify laundering or showering as energy practices would give more importance to energy than it probably deserves, and is unhelpful in attempting to distinguish between an ontology of everyday practice and the smart ontology, where energy (data and technology) takes centre stage (see Chapter 2). In contrast, I wish to emphasise that practices-that-use-energy can often become practices-that-do-not-use-energy when meanings, competencies and materials change.

Further to these clarifications, I avoid the term 'energy consumers' in the following chapters, unless I am referring to a specific construction of people as consumers, such as Resource Man. Instead, I refer to 'people-who-use-energy', 'householders', or the 'participants', 'performers', 'practitioners' or 'carriers' of everyday practices. Again, my aim is to distinguish between people's assumed primary relationship to energy as individual consumers of a commodity or users of a metered resource and their role as participants in practices-that-use-energy. While this language is slightly clumsy, it is necessary to stay away from the dominant consumer conceptualisations outlined in Chapter 3 and to mark the distinctions between the two ontologies outlined in Part I of this book.

The task ahead

In Part II I critically interrogate four common strategies of the Smart Utopia, drawing on empirical research to investigate what realities they are performing in households. In embarking on this agenda, there are several significant methodological questions that remain. First, how can we understand the ways in which the Smart Utopia is being performed through people's everyday practices? And second, where will we find the data to answer this question? The first question is one that continually plagues researchers who draw on theories of social practice to study phenomena that are conventionally oriented towards individual and rational accounts of social change. This question also relates to a bigger concern in the social sciences about how researchers can study practices without resorting to individual accounts of them. Interviews, for example, may tempt the researcher towards developing and describing what appear to be very individual and personally differentiated accounts of practices. Similarly, people may be unable to talk about their practices, given that much of their knowledge is tacit, habitual or unconscious (Hitchings 2012). This has led some researchers towards ethnography as a methodology for understanding the lived experiences of their research participants (Bell *et al.* 2005; Pink 2012a; Pink *et al.* 2010). In particular, sensory or visual ethnography, including conducting video tours, autoethnographies (diaries, for example) and following material artefacts around the home (Pink 2007, 2012a), are valuable methods for understanding the mundane and routine reproduction of practice.

My own research and the research of my colleagues (Maller 2011; Maller *et al.* 2011; Strengers 2010, 2011c; Strengers & Maller 2011, 2012) have prioritised qualitative methods of semi-structured interviews, household tours, observation and, more recently, the use of a 'memory scrapbook' (Wyche *et al.* 2006) to uncover the dynamics of household practice. Like Hitchings (2012: 65), we find that people are 'entirely able to talk about relatively mundane actions' when asked. Using the literary technique of 'defamiliarization' in the context of interviews, we attempt to position a normally mundane and seemingly inconsequential practice (such as daily showering) as unfamiliar. Bell *et al.* (2005: 153) discuss how this involves encouraging 'the participant to talk about it [any given practice] as if s/he were talking to someone from Mars'. Such techniques render seemingly mundane practices strange and contentious, allowing participants to reflect on their own actions and enabling researchers to further analyse these 'strange' experiences.

While my colleagues will continue to debate the value or otherwise of various methods used to understand and represent practice, these distinctions seem inconsequential when considering the second question posed above, concerning the origins of the data that might help us answer the first. As already evidenced in Chapter 3, the Smart Utopia is producing a vast amount of data on how Resource Man manages energy consumption and energy technologies; however, there is very little data available that has approached these issues from an everyday practice perspective. Indeed, with the emphasis on choice-based surveys nearly universally reproducing understandings and opinions of energy as a resource, commodity or impact, the lived experience of the Smart Utopia, in any form, is almost entirely absent. How, then, can we represent and understand smart energy technology's integrations into practice? On the one hand, we can't. On the other, if we read between the lines of existing research oriented towards a Resource Man agenda, the integration of smart energy technologies into everyday practices begins to emerge. Further insights can be drawn from my own research on this subject, as well as from related anthropological, sociological and human-computer interaction (HCI) design research, which has broadly sought to understand the role of new digital media and energy technologies in the home.

This is by no means a perfect science, and indeed the lack of empirical qualitative research on this subject is a critical gap in need of further attention. Nonetheless, what follows is a first and necessary attempt to understand how the strategies and technologies of the Smart Utopia are being performed in everyday life, and what this means for how energy and its technologies are understood, handled and incorporated into everyday practice. Delving into these 'dumb' and 'messy' domains of everyday life, I find significant order and intelligibility, though not always of the 'rational' kind. Further, I find a multitude of ways in which energies and their associated technologies are being integrated into practices in dynamic and diverse ways.

Part II

5
Energy Feedback

This chapter marks a shift in focus by embarking on an analysis of the first of four strategies central to achieving the aims of the Smart Utopia. In this chapter and those that follow I aim to understand how the smart ontology on which these strategies are founded is being performed by householders through their everyday practices. Further, I seek to understand what other realities are being performed as householders integrate smart energy technologies into their everyday lives. The first of the strategies discussed here is the provision of energy feedback, which is primarily intended to save energy consumption in the home.

In the Smart Utopia, energy feedback is envisioned as a key means of harnessing the interest of, and imparting the necessary energy knowledge and skills to, Resource Man in order to ensure that he is in control of his consumption (see Chapter 3). This strategy goes by various names, such as eco-feedback, home energy management, informative billing, energy monitoring, home occupancy feedback and smart energy information. What form this feedback takes, and how it is provided, varies greatly. Wood and Newborough (2007) identify over 5000 possible feedback combinations arising from energy display systems, based on where the system is located (mobile, embedded into appliances, static), motivational factors (such as competitions, goal setting and monetary rewards), units displayed (kilowatt hours [Kwh], dollars, carbon dioxide [CO_2] emissions), display methods (charts, graphs, diagrams, numbers), timescale (ranging from real-time to yearly) and category (activity, person, fuel, room). Other researchers differentiate feedback as direct or indirect (Challis 2004; Darby 2006), ambient and aesthetic (Pierce &

Paulos 2012a), or provided at the appliance or at whole-of-household level (Weiss *et al.* 2009).

Despite this diversity, the effectiveness of feedback has remained relatively constant for almost 50 years. Reviews of feedback cite studies dating back to the 1970s and consistently indicate that this strategy achieves energy savings of between 5 and 15 per cent (Darby 2006; Faruqui *et al.* 2009b; Fischer 2008). Smart metering cost-benefit analyses and large-scale energy feedback trials report more modest energy savings of 0–5 per cent (Allcott 2009; Klopfert & Wallenborn 2011; NERA 2008a). Despite over 5000 potential combinations, there is something strikingly similar about the array of feedback possibilities. This similarity can by understood by paying close attention to the unified premise of feedback that is central to the smart ontology: namely, by providing individual consumers of energy with better information or data about their energy use, they will be better able to manage and reduce it.

My aim in this chapter is to interrogate this premise by analysing how energy feedback is incorporated into or rejected from the practices of domestic life. More broadly, I am interested in what energy feedback seeks to perform and is actually performing in the home. I avoid discussing various combinations of feedback with other smart energy strategies, such as variable pricing (Chapter 6), home automation (Chapter 7) or micro-generation (Chapter 8). Instead I focus on energy feedback as a cohesive strategy in its own right.

I begin by taking stock of the different ways in which energy feedback is being delivered through smart technologies and the assumptions that underpin its provision. I find that this strategy involves householders in a limited suite of energy-saving actions. These might include changing light globes (but not the ways in which lighting is used and maintained), switching to cold water in the laundry (but not the ways in which laundry is done), or turning down the air-conditioning thermostat (but not the ways in which spaces and bodies are cooled). In defining what 'energy action' means in the context of the home, as well as what it does not mean, I suggest that many taken-for-granted practices are excluded and potentially legitimised as normal and non-negotiable activity through this strategy.

In searching for other ways in which energy feedback potentially intersects with practice, I turn to international qualitative research on energy feedback that has explicitly sought to understand its impact on everyday practice. Here I catch sight of the smart utopian vision for Resource Man, who is rationally weighing up the costs and benefits of his consumption based on the feedback provided. However, this householder is not always in control of the practices-that-use-energy in the home, making his ability to save energy somewhat limited. Instead, I find that energy feedback can be interpreted as a form of social feedback, where it provides a normative benchmark for wasteful or acceptable energy consumption across a range of practices-that-use-energy. Aside from that, energy feedback appears to play a limited role in reorienting many practices deemed non-negotiable in the home.

This does not lead me to conclude that practices are inherently non-negotiable or immovable, or that feedback is not integrally involved in changing practice. Instead, I suggest that energy feedback provides a limited set of possibilities for transforming or renegotiating everyday practice because of its explicit focus *on energy*. In contrast, I demonstrate how everyday practices are in a constant and ongoing state of negotiation in which social, material and embodied sensory feedback play a constitutive role. This presents some interesting possibilities for advocates of the Smart Utopia, which I turn to in the conclusion.

Feedback in the Smart Utopia

Energy feedback involves information provision, goal setting and/or social comparisons in relation to a metered property's (or group of properties') energy consumption. It can take a variety of forms, being delivered via the energy bill, a website portal, public installations, the energy meter, an in-home display (IHD) (also known as a smart energy monitor), text message, mobile application, or various ambient displays. Importantly, feedback doesn't have to involve a smart meter; it has been delivered in various forms for many years without the assistance of smart technology, although it has nearly always been materially mediated by some form of technology. The advent of the Smart Utopia has encouraged a proliferation of feedback devices in residential and other contexts, enabled in some cases through government mandates (see Chapter 3). Having said that, Navigant Research warns that the 'once hyped home energy management (HEM) market' has 'struggled to gain traction', and is now 'stuck in near neutral' as a result of consumer indifference and tepid utility support (Strother & Gohn 2012: 1). Despite this cautionary note, the market is now flooded with off-the-shelf IHDs such as AlertMe,[1] Current Cost,[2] ecoMeter,[3] the OWL,[4] Cent-a-meter[5] and Wattson.[6] Like other smart utopian strategies, feedback involves the extension and insertion of an explicitly technological and data-mediated ideal into the home. Traditionally used as a management tool in relation to mechanical or electronic processes (Darby 2006), 'feedback' is defined in the *Oxford Dictionary* (1989) as 'information about the result of a process or action that can be used in modification or control of a process or system...especially by noting the difference between a desired and an actual result'. Human feedback (meaning feedback intended to modify human behaviour), on the other hand, maintains this commitment to calculable, actionable information, implicating humans in their own input-output system, while acknowledging that feedback empowers individuals to make choices about their consumption. In its most simple form, the provision of information input should result in a desirable behavioural output. Resonating with the information-deficit model from behavioural science, Wilhite and Ling (1995: 150) describe this process as a 'causal link', whereby feedback is intended to result in conservation behaviour through the following linear pathway:

Increased feedback → Increase in awareness or knowledge → Changes in energy-use behavior → Decrease in consumption.

The marketing and promotion of this strategy is described somewhat differently. Here it is about putting energy consumers or more specifically, Resource Man, in control of his consumption by providing him with actionable insights (Accenture 2012a). Tendril Inc.[7] express this ambition by claiming that their products provide 'unprecedented energy insight, choice, and control' for providers and consumers. Similarly, Faruqui *et al.* (2009b: 1599; emphasis in original) argue that feedback provided through an IHD can 'turn a once opaque and static electric bill into a transparent, dynamic, and *controllable* process'.

As these examples suggest, in addition to inviting Resource Man to 'stay in control',[8] feedback is also an important method by which energy providers can *gain control* over Resource Man and his potentially unpredictable consumption. Resonating with Foucault's (1995) analysis of 'disciplinary technologies', and more recent analyses of neo-liberal government attempts to establish self-governing (and governable) citizens (Rose & Miller 2010), energy feedback constitutes householders as calculative agents who are constantly weighing up the environmental impacts of their individual actions (Marres 2011). In this way, energy utilities can extend their influence inside the home, both physically through the feedback display or device, and metaphorically by bringing their *modus operandi* – 'you can't manage what you can't measure' – into domestic life.

In order to render energy a calculable and actionable entity that can be metered, measured and managed, feedback emphasises energy's role as a commodity, resource unit or impact – that is, as something that both can and should be counted and managed. In short, feedback implies that energy is *in need of* commodification. For example, Fischer's (2008: 80) observation that energy consumption is not a 'coherent field of action' but rather a suite of activities such as cooking meals and working on the computer, leads her to conclude that energy is a 'low interest product', but a product nonetheless. Similarly, feedback studies regularly argue the benefits of this strategy by pointing out the lack of information consumers currently have available to them regarding their use of energy. For example, Kempton and Layne (1994) point out the absurdity of receiving a non-itemised monthly bill for other products, such as groceries or petrol. Other authors make links to car speedometers, which provide regular and instantaneous feedback, suggesting that energy consumption needs a similar indicator of performance (Wood & Newborough 2003). Thus, even though energy is acknowledged as being substantially different from other commodities, feedback aims to turn this immaterial material (Pierce & Paulos 2010) into a tangible and manageable product.

Despite this similar starting point, there is no agreement over the 'best' format, form or frequency of feedback. Some say web-based feedback works (Vassileva *et al.* 2012), others say it has limited efficacy. Some say the novelty of feedback wears off after time (Hargreaves *et al.* 2010, 2013) (known as the 'fallback effect' (Wilhite & Ling 1995)), others say it is sustained (Darby 2006); and there is an ever-growing list of 'other factors' that play an 'important role' in determining the

78 *Smart Energy Technologies in Everyday Life*

effectiveness of feedback (Faruqui *et al.* 2009b: 1607). Where feedback has not achieved desired or anticipated consumption reductions, some advocates contend that this is because its recipients did not have the 'right' information in a format they could use, with enough regularity to hold their interest. The frequency of feedback is often considered of critical importance: not only must Resource Man be 'properly informed of actual electricity consumption and costs', but this information must be provided *'frequently enough to enable them to regulate their own electricity consumption'* (European Commission 2009 in Darby 2010: 448; emphasis in original). While researchers continue to debate these differences and distinctions, there is now a growing body of evidence to suggest that energy feedback focuses householders on a narrow range of energy-saving actions that define what energy-saving is, and more problematically, what it is not.

Performing energy-saving actions

One of the ways in which energy feedback 'works' is by recruiting householders into a coherent package of energy-saving actions. This is reflected in a growing number of feedback studies which report surprisingly similar findings. For example, Anderson and White (2009: 10) classify the direct actions taken in relation to appliances by their participants in a trial of different IHDs as follows:

- Turn it off.
- Use it less.
- Use it more carefully.
- Improve its performance.
- Replace it/use an alternative.

Coming from the disciplinary perspective of human–computer interaction (HCI) design, Pierce and Paulos (2010b) propose a similar 'vocabulary' to describe the energy conservation actions undertaken in response to energy feedback. These involve cutting power by turning an appliance off, trimming energy use by using a 'low' or more efficient setting, switching or upgrading to a more energy-efficient product, and shifting usage to a different time or place, without necessarily reducing total energy use (Pierce *et al.* 2010b: 1987).

The findings from these and other studies suggest that feedback encourages 'small changes'. Indeed, the promotion of energy-saving 'tips' alongside the provision of feedback often makes this an explicit aim of many programmes (Fischer 2008). There are now familiar lists of energy-saving actions promoted on government and energy utility websites and reported in statistics. Sometimes referred to as 'ten easy actions', these lists seek to define what it means to save energy and reduce the effects of climate change and/or peak electricity demand.

In some ways, we could think of these 'actions' as a bundle of practices involving specific skills (about how to understand and use energy data), materials (home automation, timers, IHDs, smart meters etc) and meanings (rationality, efficiency and budgeting) that seek to constitute, and place boundaries around, what it means to manage and save energy. In other ways, we could think of these actions as specific energy-saving 'elements' of practice that attempt to reorient practices of laundering, cleaning or cooling by bringing in new skills, meanings and materials. However, while encouraging householders to turn their thermostat up or down to save energy and reduce CO_2 emissions might bring meanings of rationality and efficiency to the practice of cooling a home, it is unlikely to fundamentally change how cooling is done or what materials are required. For this reason, energy feedback might be best thought of as enabling a limited suite of 'actions' that constitute 'moments' in a practice (such as turning down the thermostat) rather than fundamentally reconfiguring practice itself.

Framing activities that arise from the provision of energy feedback as a tightly bundled set of energy-saving actions raises a number of possibilities in regard to how energy-saving is performed, and indeed *what* is being performed. Marres (2011) has previously examined the performative capabilities of different carbon-accounting devices in this way. She characterises energy-saving actions in terms of the 'work' required to undertake them, arguing that participation in these actions amounts to a 'change of no change', whereby devices of carbon accounting facilitate a form of public participation that involves minimal or no effort. Similarly, Hobson (2011: 200) argues that participation in a limited suite of energy-saving practices allows householders to 'rest easy' knowing that the planet is being saved, while other practices-that-use-energy escalate in their resource consumption. In this way, participation in energy-saving practices is a means by which householders can relate to largely intangible and seemingly unsolvable environmental and energy management issues, while doing very little at all. Here, IHDs and HEM systems are positioned as the 'material and physical linkages' (Marres 2010: 182) or the 'mediators of public involvement' (Marres 2010: 179) between everyday practices and national and global energy problems. In this way, feedback systems are part of a wider suite of strategies that render it possible for the simple act of turning off the light to allow householders to perform the seemingly insurmountable task of saving the planet.

In other instances, the provision of feedback (and other forms of energy or carbon accounting) frames participation in energy-saving practices as 'hard work', involving regular reflection, conscious deliberation and disruptive activities such as installing insulation or undertaking a major house renovation (Marres 2011). 'Green' renovations are positioned at the extreme end of this 'work', where householders participate in saving the planet by installing 'big ticket items' such as more energy-efficient appliances and ceiling insulation. However, like the 'easy' energy-saving practices facilitated by energy feedback, Maller *et al.* (2011: 18) observe that 'green' renovations can at times appear 'paradoxical': householders attempt to green their home while also making modifications that will counteract these attempts, by increasing the floor area, adding bathrooms, or extending kitchens and living areas. This leads these researchers to conclude that home renovation practices run up against 'existing or future daily routines and aspirations for the ideal home' (Maller *et al.* 2011: 18). Their analysis suggests that even the hardest of environmental labour, or the most difficult energy-saving practices, can obscure and exclude many of the changing practices of everyday life.

The continuing problem here is that energy-saving actions, big and small, not only define what energy-saving is, but also what it is not. The feedback expert Darby (2008: 502) alludes to this problem when she warns that 'savings' achieved through energy feedback programs 'so often turn out to be steps taken down an upward-moving escalator'. Thus, even when energy savings are maintained over time through the continual performance of energy-saving actions, these savings can be negated by householders' participation in *new* practices-that-use energy, or the changing configuration of existing ones. The emergence of outdoor heating (Hitchings 2007), more frequent laundering and showering (Davidson 2008; Slob & Verbeek 2006) and cooling practices involving the air-conditioner (Ackermann 2002; Shove 2003) are all examples of the continual change in what counts as normal, taken-for-granted practice. In the case of feedback, 'what counts is often what can be relatively easily counted', with energy being revealed in a limited suite of energy-saving practices, while other practices-that-use-energy are moved into the shadows (Shove 1997: 270).

Energy feedback in everyday practice

In order to consider what else energy feedback might (or might not) be performing, it is helpful to consider some examples of studies that have explicitly sought to understand the role of energy feedback in everyday life. Three qualitative researchers, including myself in association with colleagues, have studied these in detail (Hargreaves 2010; Hargreaves *et al.* 2010, 2013; Pierce *et al.* 2010a; Pierce *et al.* 2010b; Strengers 2011b, 2011c). All have conducted studies with householders using feedback systems in small samples across three continents (UK, US and Australia). Our findings are remarkably similar, and illustrate the limitations of energy feedback in changing practices-that-use-energy because of this strategy's explicit focus on energy. While these studies independently found that householders are recruited into and continue to perform the energy-saving actions discussed above, they also found that energy feedback is limited by seemingly non-negotiable practices which vary substantially between households. I begin this discussion by drawing on these studies to locate Resource Man, noting that where he does exist, he lacks control over the practices-that-use-energy in the home.

Locating Resource Man

Studies of energy feedback conducted by myself (Australia), Pierce and colleagues (US) and Hargreaves and colleagues (UK) have found little evidence of the rational Resource Man central to the Smart Utopia (see Chapter 3). The first indication of his absence is that energy management terminology, such as kWh, CO_2 emissions and even in some cases dollars and cents, were somewhat irrelevant terms for the participants of these studies. Householders referred to kilowatt hours as 'kilowattevers' or 'googalldygook' (Strengers 2011b: 2138, 7) and found CO_2 emissions 'meaningless' (Hargreaves *et al.* 2010: 6114; also reported in Anderson & White 2009: 25). Pierce *et al.* report considerable indifference in learning more about the cost of energy, even in the most actively engaged households. For example, one participant commented 'I know I'm not gonna change anyway, so I don't really wanna know' in relation to information about the costs of their energy consumption (Pierce *et al.* 2010b: 1988; participant quote). However, perhaps the most striking indication of Resource Man's absence, is that in one of Pierce *et al.*'s (2010a) studies, not one participant even used their free interactive energy monitor – not even to test it out.

This was not the case in all of these studies. Where energy data and information was welcome and appealing, it was of more interest to the (resource) man of the house (Hargreaves *et al.* 2010, 2013; Strengers 2011b). This was evident early on in my research when I was trying to recruit entire households to participate in the study. Some women agreed to participate reluctantly and only by specific request, later commenting that this was because they had limited interest in and understanding of either energy consumption or the IHD supplied to their home: 'I don't know how to use it. It's got nothing to do with me' (Strengers 2011b: 2140; female participant quote). Similarly, in Hargreaves *et al.*'s study there was often a single dominant user of the energy display, and this person was more likely to be male:

> D3 [Participant]: I must admit it's mainly blokes [who've shown an interest in it].
> I [Interviewer]: Why do you think that is?
> D3: Oh, we just like flashing lights and fiddling with knobs and things, don't we? (Hargreaves *et al.* 2010: 6115)

Hargreaves *et al.* (2010) do note that this was not the case for everyone, and that some women and children were interested in the display, although teenagers rarely were. This finding is consistent with my own research, and with studies of consumption more broadly (*e.g.* Gram-Hanssen 2007). However, where young children were interested in energy feedback and changing certain practices they considered unnecessary or wasteful, they were sometimes prevented from doing so by their mothers, as the following dialogue indicates:

> SON: No-one uses the hair dryer anymore.
> DAUGHTER: Mum forces me to ... It's because she doesn't like the way it looks if I don't blow dry it.
> MOTHER: [Laughter] It takes about two seconds. I'm not going to stop because of that [referring to IHD]. (Strengers 2011b: 2140)

These findings are unsurprising when we consider the domestic history of the home, which has been, and in many ways still is a female domain (Carlsson-Kanyama & Lindén 2007; Schwartz Cowan 1989). Women are still responsible for the majority of household labour in many Western countries (Carlsson-Kanyama & Lindén 2007), particularly in regard to the comfort, cleanliness and care of their home and family. These findings suggest that feedback may be of more interest to members of the household who have little control over, influence or interest in many practices-that-use-energy.

In a revealing example from my research, a stereotypical Resource Man (engineer by profession) was reportedly counting the kilowatt hours (and water consumption) of his home every day by accessing a website portal at work. However, his efforts to reduce and manage energy and water were confounded by his wife, who was at home undertaking domestic practices, such as laundering and house cleaning. She described her husband's monitoring as 'sort of funny' and 'very Big Brotherish', but did not substantially change her practices-that-use-energy or water as a result (Strengers 2009: 158).

In other examples, feedback systems were used as an energy-management tool in line with the aspirations intended for Resource Man and similar to a speedometer on a car, giving an overall indication of a household's energy performance and providing a normative benchmark (Hargreaves 2010; Wood & Newborough 2003):

> It's like the speedo on a car. Years ago, people would drive at whatever speed they wanted to. But now we understand that there's a limit. (Strengers 2009: 149)

This was not the case for all householders in my study. Some of them rejected the Resource Man role intended for this device: 'That's how I drive a car but it's not how I'd live at home' (Strengers 2009: 153). Where householders did embrace the speedometer attributes of energy feedback, they mainly used it to develop an understanding of the 'limits' of their energy consumption, rather than to make rational decisions. Hargreaves (2010: 27) reports similar findings, noting how one participant described a period of high consumption as 'using it hot', resulting in this householder taking immediate action to reduce the household's energy usage. Hargreaves (2010: 28) refers to this type of response as the 'nag factor' of feedback, and comments that it results in the relatively small energy-saving actions discussed earlier, such as turning off the lights or standby power. This nag factor also points towards energy feedback being positioned as a new form of social feedback, providing normative benchmarks about how or *how much* energy could and should be used to undertake a specific practice or suite of practices.

Social feedback about energy

Energy feedback can monitor a household's energy consumption, providing normative indications about whether a household's practices are 'wasteful' or 'acceptable'. There are a number of ways in which this social feedback is provided. In my studies, it was conveyed to households through a series of 'traffic lights' displayed on their IHD, which indicated high (red), medium (orange) and low (green) periods of consumption, or peak (red), shoulder (orange) and off-peak (green) pricing periods. One of my participants used the term 'redlining' (Strengers 2011c: 328) to refer to a period when their IHD displayed a red light (indicating high consumption in this case), during which time this household did everything they could to bring their consumption back down to green. This involved temporarily suspending many practices-that-use-energy and rescheduling or reordering them for a short period of time. This outcome is very close to the desired load-shifting objectives of energy demand managers: the idea of energy as a limited material was incorporated into a range of practices-that-use-energy for a short period of time, with the aim of smoothing out any spikes in consumption.

However, just as a red light can convey meanings of excess and waste, a green light can be interpreted as just that – a 'green light' to consume more. For example, some of the households involved in my research indicated that the display of a green or even an orange light during a specific activity, such as using the dishwasher or clothes dryer, served to legitimise those specific appliances or practices. One of my research participants summed this up as follows: 'I was always worried about using the dryer so much, but I figure it doesn't make it scream red so it's OK' (Strengers 2011b: 2140). Similarly, other feedback studies (Pierce *et al.* 2010b; Reidy *et al.* 2005) have found that once householders see the low cost of running specific appliances, they may use this as a justification for their continued usage.

Another form of social feedback is the establishment of a 'baseline' or 'normal' level of consumption, which householders try not to exceed. Once householders understand their baseline, which is measured in relation to their own consumption, householders have little use for energy feedback, aside from using it every now and again to ensure that they don't drift too far above this benchmark (Hargreaves *et al.* 2013; Strengers 2011c). This is one explanation for why the 'novelty effect' of feedback might wear off over time. Hargreaves *et al.* (2013) warn that the idea of a baseline can become a self-fulfilling prophecy, with householders reluctant to reduce their consumption below their 'normal' level, instead attempting to justify and defend 'baseline' activities as 'necessities' that should be taken for granted.

A more specific example of establishing normal benchmarks can be found in feedback based on 'average' consumption or pre-determined standards. Seligman *et al.* (1978) and Intille's (2002) studies provided feedback to householders to indicate when they could cool their houses by opening a window. In both of these studies, specific temperatures were used to recommend when opening a window would be preferable to running an air-conditioner during summer. Seligman alerted householders to the presence of the window at an outdoor temperature of 68°F (20°C) when their air-conditioner was still running, whereas Intille's hypothetical scenario proposed opening a window at an indoor temperature of 74°F (23°C) and an outside temperature of 78°F (25.5°C). Seligman *et al.*'s study achieved a 15.7 per cent reduction in energy consumption, which sounds impressive. However, this strategy may have also introduced a new normative meaning to the practice of cooling the home, by reinforcing the understanding that air-conditioning is a necessity at temperatures above 20°C. A similar warning is repeated by adaptive thermal comfort researchers, who continue to argue that the establishment of homogeneous and climate-controlled temperature standards in indoor buildings can normalise and potentially legitimise the necessity of air-conditioning (Brager & de Dear 2003; Brager *et al.* 2004; Chappells & Shove 2005; Strengers 2008). In these and other ways, normative meanings conveyed by social feedback either intentionally or inadvertently inform existing everyday practices, rendering them excessive and wasteful, or normal and necessary. One of the potential outcomes of this process is that normal and necessary practices are positioned as non-negotiable, at least in terms of the energy they require to be performed. Energy feedback strategies can attempt to render these non-negotiable practices negotiable by bringing these activities into conscious reflection and analysis. However, these attempts have so far had limited effect because they assume that negotiability is a conscious decision-making process.

Non-negotiable practices

Even when feedback is correctly interpreted and where energy-saving actions are being performed, it is often considered irrelevant in the context of the home, where many practices are considered non-negotiable or inflexible (Pierce *et al.* 2010b; Strengers 2011c). In one of Pierce *et al.*'s (2010a: 247) studies, a research participant commented: 'I'm not gonna not wash my clothes!', resonating with the finding reported more broadly (see Anderson & White 2009; Darby 2010) that people often feel that they are already doing all they can to save energy. Similarly, in my research on energy feedback, a participant made it clear that energy feedback would not result in the negotiation of many of their daily practices:

> It might be nice to know that the toaster is this and the kettle is this, but I don't know what I'm supposed to do about it – have cold tea? (Strengers 2011c: 331)

One of Hargreaves *et al.*'s (2010) research participants also commented that many appliances are considered 'necessities' that cannot be compromised, or should not be comprised if one wants to enjoy a 'good life':

> There are some things you just can't change. So, as I say, I have my fish tank and the fish need a pump, and I cook, so I can't really

change that. I mean, I think that life is for living and I don't want to become obsessive about it or like Scrooge or anything. I want to enjoy living and working in my house. (Hargreaves *et al.* 2010: 6117)

There is great diversity in what activity is reported as non-negotiable, with householders identifying a range of appliances, such as computers, air-conditioners, televisions, tumble dryers, breadmakers, Venetian lamps and fish tanks, as unable to be compromised or changed (Hargreaves *et al.* 2010; Strengers 2009). Feedback and behavioural studies grounded in the smart ontology refer to these seemingly diverse and idiosyncratic differences as 'user preferences', which are thought to result from variations in personal histories, attitudes, socio-cultural demographics (age, gender, education, income), physical or mental health, relationships, and available free time (Wood & Newborough 2003). These preferences retain their persistent characteristics in the form of habit, which Triandis' Theory of Interpersonal Behaviour defines as 'situation-behaviour sequences that are or have become automatic' (Triandis 1980 in Darnton *et al.* 2011: 23). Individuals are not usually 'conscious' of these sequences, and therefore behavioural cues and goal setting are thought to be needed to bring habits back under cognitive control (Wood & Newborough 2007). The goal of feedback is to draw each individual's attention to their habits, so that activities deemed non-negotiable can be rendered negotiable once more. However, this does not always eventuate. Indeed, as my own research and the studies discussed in this chapter demonstrate, feedback only renders a very specific suite of energy-saving actions negotiable, while potentially reinforcing and legitimising many others as non-negotiable.

From this we could simply conclude that many practices are immovable and leave the discussion at that. However, other studies find that practices are continuously negotiated, just not in the ways outlined above. More specifically, through different forms of *social*, *material* and *embodied sensory* feedback, practices are constantly being negotiated and renegotiated in ways that are both self-sustaining and potentially transformative. In order to understand why energy feedback is potentially limited in its ability to renegotiate practice, we first need to take stock of the different ways in which other types of feedback operate within and orient everyday practice.

Negotiating everyday practice

Shove et al. (2012: 99) argue that 'monitoring, whether instant or delayed, provides practitioners with feedback on the outcomes and qualities of past performances'. This feedback potentially 'feeds forward' into what these practitioners do next, thereby playing an important role in 'the persistence, transformation and decay of the practices concerned' (Shove et al. 2012: 99). Feedback can help determine what it means to do a practice well, or what it means do a practice at all. Furthermore, data can play an integral role in this process. For example, Shove et al. (2012) draw attention to the growing range of instruments of recording (the body, the score sheet) which now play a constitutive role in many sporting practices. They argue that devices such as heart rate monitors and pedometers have 'modified the meaning and experience of physical exercise for millions of people' (Shove et al. 2012: 100). Practices are being renegotiated.

Similarly, the growing popularity of Garmin[9] fitness technology is enabling the collection of an unprecedented quantity of actionable data that in turn can be shared and compared with other data through a range of associated technologies. This data is subtly changing a range of fitness practices, where it is used to engage participants in continual forms of self-improvement. Social networking fitness platforms, such as Strava[10] and Map My Run[11] are transforming common cycling and running training routes into racing circuits, where participants are engaged in bettering their own data as well as the data of their friends, colleagues or anonymous virtual 'racers'. In these and other ways, data *feeds back* to participants how fit they are, where they can improve, and how much more training they need to do.

You would not be wrong in thinking that this description of data and technology sounds incredibly similar to the aims and intentions for energy feedback in the Smart Utopia. Why then is one form of data (fitness data) able to renegotiate practice where another (energy data) seemingly does not? One plausible explanation is that monitoring and measuring performance in quantitative terms, such as time, speed and distance, has long been a feature of fitness practices. Racing, time trialling and record setting are oriented towards producing and acting on data: here the meanings, skills and materials of calculating and establishing benchmarks are absolutely essential to the performance of the practice.

On the other hand, feedback about energy is not currently integral to many practices of domestic living, except perhaps in low-income families, where the high cost of energy relative to other living expenses renders it an important and meaningful element of practice that is already closely monitored (Dillahunt *et al.* 2009). This does not mean that feedback is not still an essential aspect of these practices. Rather, other forms of social, material and embodied sensory feedback may take precedence over energy feedback. For example, social feedback might involve our friends giving us a tip on how to get a tough stain out of a piece of clothing, or our partner complaining that the room is too cold. Even pets and other non-humans might monitor practices by sitting suggestively next to the heater on a cold day, providing a subtle form of feedback that encourages their owners to turn the heater on.

Other forms of feedback are explicitly material. For example, the home and its surrounds might 'invite' its occupants to open its doors and windows to capture a fresh breeze depending on its design and orientation, or it might encourage the provision of cooled and dehumidified air by prioritising thermostatic controls over openable windows and doors (Brager *et al.* 2004). Similarly, devices and technologies in the home might provide us with feedback about what cycles we should put the washing on, or when we should maintain and clean them. Sofoulis (2005: 458) makes a similar argument when she describes showers and washing machines as 'saver-unfriendly obstacles... designed for using and wasting water, not for conserving, reclaiming and reusing it'. They embody a 'fantasy of endless supply' (Sofoulis 2005: 452), providing 'feedback' that using more energy and water is expected and accepted, while using less is inconvenient as well as labour- and time-intensive. In this example, the provision of energy (or water) feedback potentially competes with the material feedback literally built into the design of the shower.

Our bodies are also important self-monitoring devices, providing feedback through our senses, such as whether laundry, bodies, rooms or last night's dinner looks or smells dirty or 'off', or how clothes and linen 'feel'. These visual, olfactory and tactile 'ways of knowing' have been a feature of Pink's (2004, 2005, 2012b) ethnographic work on the home as a sensory environment, where they also intersect with the sensory perceptions of others (*e.g.* how 'others' think clothes should look, feel and smell), constituting hybrid forms of social and sensory feedback. Our bodies provide other forms of feedback too, by letting us know when we feel cold, and by constituting an important 'testing' instrument for establishing comfortable temperatures both inside and outside the home.

Feedback about energy both interacts and potentially competes with these other forms of social, material and embodied sensory feedback. Like energy feedback, we can think of these as having different frequencies and timescales at which they operate and become important to practice. Recognising that other forms of feedback exist does not preclude the possibility that data about energy can become an integral form of feedback, but it does suggest that this data is not currently essential to many practices-that-use-energy.

Consider, for example, Hand and Shove's (2007) research on 'living with a freezer', which clearly demonstrates the multiplicity, creativity and continual negotiation of living with this seemingly 'essential' appliance. Freezing, these researchers argue, involves 'the active and simultaneous integration of constituent elements'. Materials (food, freezers, kitchens, Tupperware), meanings (of healthy eating, economy, proper provisioning, 'good' food) and skills (in planning, cooking and shopping), come together in a 'provisional and inherently unstable arrangement' which changes as these elements change (Hand & Shove 2007: 96–7). We can think of feedback as being involved in this process, being constituted as recommendations on food packets about where they should be kept, for how long and at what temperature; the labeling of freezer bags and containers to 'remind' householders when or how to use them; tasting and smelling food to 'check' if it is still OK to eat; and menu plans stuck to the fridge or freezer. In this context, the delivery of energy feedback makes little attempt to intersect with or renegotiate the elements of freezing as a practice.

Importantly, while the social, material and embodied sensory feedback described above might be integral to the performance of everyday practices, it is not always or necessarily involved in a practice's transformation. What householders actually do in response to feedback depends on the elements (materials, meanings, skills) in circulation, or on the way in which the practice is currently constituted (Shove et al. 2012). Advertising and marketing companies recognise this when they seek to integrate new meanings, skills and materials (e.g. products) into existing practices by suggesting what people should do in response to social or sensory feedback. For example, marketing companies provide recommendations about what people should do when they smell odours in their homes or on their bodies, or when they see dirty marks on the bathroom walls, toilet or laundry. While these attempts do not always lead to change, the point I wish to make is this: for feedback, of any type, to facilitate the changing constitution of a practice (e.g. to change what a practice actually is and how it is performed), it must also be involved in changing what makes sense for people to do (Schatzki 2002).

As a final example, let us consider energy feedback that is directly targeted towards specific household activities or practices (Seligman et al. 1978; Weiss et al. 2009; Wood & Newborough 2003). Can energy feedback, such as the commercially available Kill A Watt[12] device that can meter any individual appliance, play a constitutive role in the renegotiation of practice? In one study, the provision of energy feedback was directed to the practices of cooking, resulting in new energy-saving skills, meanings and materials being integrated into these practices, such as replacing a more energy-intensive appliance (oven) with a less energy-intensive one (microwave), reducing hob (stovetop) and oven cooking times, reducing the amount of water in pans or the kettle, putting lids on pans, and/or simmering water instead of boiling it (Wood & Newborough 2003). Notice, however, that this feedback still retains its explicit focus *on energy* (as a commodity, resource unit and impact). New energy competencies, meanings and materials are pitted against the other elements of cooking practices (such as how to prepare a healthy, quick meal or how to host a dinner party) – elements which are constantly being renegotiated as people's schedules and life courses change (Hand & Shove 2007), and as cooks become recruited into new practices (Truninger 2011). As with other forms of feedback, a continued explicit focus on energy reduces the potential for 'savings' to be maintained amidst the continually transforming elements of cooking.

From this brief discussion we can conclude that practices *are* negotiable, although not only or often because of the provision of expert information about energy as a technical and measurable commodity. Rather, other forms of social, material and embodied sensory feedback are integral to the ongoing performance and transformation of practice. In focusing *only* on energy feedback, other forms of feedback remain largely hidden, as do the many other ways in which practices are continually negotiated, potentially limiting the Smart Utopia's ambitions of reducing and shifting energy consumption.

Feedback that matters

I have argued that energy feedback is one among many forms of feedback that are integral to everyday practice. Drawing on international qualitative research, I suggested that even where energy feedback performs the vision intended for Resource Man, it has a limited impact on practices-that-use-energy. More specifically, it encourages householders to participate in a small suite of energy-saving actions that result in the 'change of no change', even when 'hard work' is involved (Marres 2011: 527). Further, energy feedback can provide a form of social feedback that brings new normative meanings to how much or how little energy various practices ought to be consuming. Beyond this, energy feedback currently appears to have a negligible impact on practice: many domestic activities remain non-negotiable. Rather than seeing this as a fundamental limitation of all programs aimed at reducing energy consumption, or as indicating the emergence of a broader environmental inertia, I have argued that these observations can be explained by the limited assumptions on which the delivery of energy feedback is based and on the continued focus of energy feedback *on energy*.

The consistent finding that householders report a diversity of non-negotiable practices should be interpreted neither as a fixed state of affairs nor as a reason to give up on attempting to facilitate change. Practices are always 'on the move', as are their elements (Shove 2003: 2), and it is this continual movement that presents opportunities for change. Social, material and embodied feedback play a critical role in the transformation of domestic practices-that-use-energy, even if energy feedback appears to play a limited one. From this we can conclude that feedback is important to the transformation of practices-that-use-energy, even if *energy* feedback is not.

This does not mean that energy feedback will never matter to practice, or that it will only ever matter to practice in restricted ways. But it seems at least plausible to suggest that in order for energy feedback to matter, energy itself must matter. Energy must become important to what it means to do the laundry, cook a meal or entertain a guest. This might be one line of enquiry that advocates of the Smart Utopia could pursue in order to develop smart strategies that matter to everyday practice. Another might be to put energy to one side, and instead focus on the many other forms of feedback that are integral to everyday practice. Advertisers and marketers are already involved in this process: they provide explicit forms of feedback about how to produce clean and fresh clothes, homes and bodies. Similarly, house designers, builders and appliance manufacturers are implicitly and explicitly involved in designing and making things that provide material feedback.

At best, energy feedback encourages participation in a small suite of energy-saving practices and provides social benchmarks that may in time lead to larger savings. At worst, it obscures and legitimises broader changes in 'normal' practice not deemed accountable as energy-saving activity, as well as masking other forms of social, material and sensory feedback that are an integral part of the ongoing performance and transformation of everyday practice. In the following chapter I turn to the smart strategy of dynamic pricing, arguing that it has the potential to transform the meanings of energy – to make energy matter – albeit for specific periods of time.

6
Dynamic Pricing

The success story of dynamic or variable pricing is told by economists. Evidence from 30 years of sustained economic research demonstrates substantial and sustained reductions in household electricity demand during peak times – between 13 and 20 per cent for critical peak pricing (CPP) – or much more (27–44 per cent) when combined with energy feedback and 'enabling' technologies such as smart thermostats (Faruqui & Sergici 2010). These broad-brush statistics mask the ways in which dynamic pricing intersects with everyday routines and what becomes malleable and adaptable as a result. Analysed almost solely in terms of householders' price responsiveness, electricity consumption is positioned as relatively inelastic, inflexible and non-negotiable unless 'the price is right'. However, this position is continually confounded by the finding that the seemingly non-negotiable practice of air-conditioned cooling is *increasingly* negotiable with hotter temperatures and where more consumers own these devices (eMeter 2010; Faruqui & George 2005).

In seeking to understand these apparent anomalies this chapter steps sideways from the rising tide of statistics on dynamic pricing, grounded in the smart ontology. Focusing specifically on CPP, the chapter reveals the elasticity of many practices-that-use-energy, particularly those involving air-conditioned cooling, during network peaks. Analysing this issue through the lens of everyday practice (see Chapter 4), I contend that CPP encourages the circulation of new meanings about energy as a valuable, scarce and rare material during the 'exceptional' period of peak demand. In contrast to the strategy of energy feedback discussed in Chapter 5, energy comes to matter during a CPP 'event'. More specifically, CPP can shift the meaning of air-conditioned cooling from a necessary to a wasteful activity, and encourage the resurrection of a range of alternative cooling practices that don't involve the use of this device. This disruption is not unusual or unnatural; rather, everyday life is full of breakdowns, disruptions and 'special circumstances' that encourage flexibility, innovation and adaptation in practice. What's more, householders have never had more (electrically enabled) tools to help them renegotiate practices-that-use-energy in the home. The chapter begins with a brief summary of dynamic tariffs and peak pricing in the residential sector, where research has been framed within the smart ontology. Here a number of questions remain about how and why dynamic pricing, and more specifically CPP, 'works'.

Pricing the peaks

Energy analysts continue to lament that for the past century, electricity pricing has not reflected its true costs (Faruqui and Sergici 2010; Sioshansi 2012). The majority of electricity provision is still charged on a flat tariff, without taking into account the time differentials in demand and their associated expenses. This is despite the fact that the overall story of dynamic pricing is a positive one. Faruqui and Palmer's (2011) review of pricing trials indicates strong support for dynamic pricing among those who have experienced it, even though they cite 'consumer inertia' when it comes to actively switching pricing plans. Faruqui et al. (2007) conclude that 80 per cent of customers would stay on dynamic tariffs if they were offered as the default rate.

With the international adoption and proliferation of electrical heating, and more recently, cooling technologies, flat pricing is becoming increasingly problematic for electricity utilities. Major network peaks, which historically occurred on the coldest days of the year, are now occurring on the hottest in some nations, as electricity systems struggle to keep up with increasing industrial, commercial and residential demand for coolth. Peaks generally occur as residential consumers return home from work – from the early afternoon into the evening, where there is overlap with industrial and commercial load. The residential sector is the fastest growing contributor to peak demand in many countries, and constitutes a significant slice of the overall peak (Faruqui & Palmer 2011).

In North America, these peaks typically occur during the top 1 per cent of hours, where 9–17 per cent of the annual peak demand is concentrated (Faruqui & Sergici 2010). In other words, 9–17 per cent of North America's electricity system (generation, transmission and distribution) sits unused for 99 per cent of the year. This is a familiar story for other nations such as Australia, which has experienced rapid growth in air-conditioning demand over the last 50 years – and demand is expected to rise further (DEWHA 2008; EES 2006). Around 15 per cent of Australia's National Electricity Market capacity is only needed for a combined total of four days of the year. That equates to an AU $11 billion investment in network equipment for around 100 hours of electricity provision (Lohman 2011). In this context many economists view dynamic pricing as an equity issue. Proponents argue that those who cannot afford air-conditioning or choose not to have it are paying for those who do (Faruqui 2012). In one estimate, householders without air-conditioning in the Australian state of New South Wales were found to be cross-subsidising their cool neighbours by AU$70 a year to pay for the extra infrastructure required to power these appliances (Frew 2006).

From the electricity industry's perspective, peak demand is essentially a problem of inefficient investment. Billions of dollars' worth of electricity assets are sitting idle for the majority of the year, and its cost is passed onto all electricity consumers through their bill. This is why many utilities are interested in load 'smoothing' and 'shifting' through the introduction of dynamic prices, and why some utilities encourage householders to cool their homes before or after peak periods. Small drops in peak demand can have big consequences. Faruqui *et al.* (2007: 74) estimate that a 5 per cent national (US) reduction in peak demand is achievable with dynamic pricing, and can eliminate the need to install and run around 625 infrequently used peaking power plants[1] and their associated delivery infrastructure. However, the aim of smoothing and shifting demand, rather than shedding it, is, for some analysts and industry commentators, in conflict with, or at least separate from, strategies to reduce greenhouse gas emissions (Hledik 2009). Indeed, the 'conservation effect' of pricing programmes (*e.g.* overall or average energy demand reduction) is much less impressive than the demand response (*e.g.* shifting demand away from peak times), and more widely debated (King & Delurey 2005).

When utilities speak about dynamic or variable pricing, they are usually referring to one of three pricing tariff structures. Time-of-use (TOU) tariffs involve two or more rates for different times of the day, generally identified as off-peak, shoulder and peak periods. Prices are lowest during the off-peak period and highest during the peak, reflecting the extra costs of supplying electricity during peak periods. TOU tariffs reflect the average peaks and troughs in daily demand, but do not reflect the price associated with providing electricity during the critical or network peaks (top 1 per cent of hours) (Borenstein *et al.* 2002).

CPP, also known as dynamic peak pricing or interruptible electricity rates, reflects the additional costs of providing electricity during these network peaks. Unlike the regularity of TOU, the CPP rate is only dispatched on a limited number of 'critical' peak demand days, which normally occur during the summer or winter months. In most instances, somewhere between 12 and 20 CPP days are called throughout any given year for a short period of time (2–5 hours), during which the price of electricity is 10–40 times the off-peak rate (Herter 2007; Strengers 2010). Consumers are usually notified of these critical peaks via a range of ICTs (SMS, email, automated phone message, IHD) approximately one day in advance. A variation on CPP is the critical peak rebate (CPR), also known as the peak time rebate or a dynamic peak rebate, which rewards customers for reducing electricity during these critical peaks with a rebate on their bill. In some trials, CPP or CPR is layered over the top of TOU tariffs, communicating the 'normal' daily peaks to customers, as well as signalling critical peak events (Faruqui *et al.* 2009a).

Finally, there is real time pricing (RTP), which is considered the most accurate cost-reflective price signal, passing on the hourly market rate of electricity direct to electricity consumers, who are alerted to the hourly prices either a day or an hour ahead. RTP is generally thought to be unsuitable for residential consumers due to the large price risks it would expose them to, and so is typically only provided to large industrial customers (Faruqui *et al.* 2009a; Herter 2007). Given that a number of researchers are now calling on the governments of many modernised nations to make dynamic pricing, particularly CPP, the default rate for residential electricity consumers (Borenstein *et al.* 2002; Faruqui & Palmer 2011), I focus on this pricing strategy for the remainder of this chapter.

There is little analysis of why CPP generates a much higher demand response than TOU, aside from the higher rates for electricity which are charged during CPP events. However, as Faruqui and Sergici (2010: 221) note, 'higher prices do not induce proportionately higher responses, confirming once again that the law of diminishing returns is at work'. Furthermore, economic analyses prove to be of little explanatory value when considering why some CPR programmes, and even some 'information-only' trials,[2] which notify households of a critical peak event but provide no financial incentive or disincentive to respond, achieve higher peak reductions than TOU tariffs. In an Australian trial of CPP, for example, the information-only group reduced their peak consumption by an average of 13 per cent in the summer, and 11 per cent overall (Strengers 2010). Households received notification of a peak event through a number of communication media. In contrast, Charles Rivers Associates' analysis of California's Statewide Pricing Pilot found that the information-only group response was small, unstable or insignificant (CRA 2005: 86).

These two contrasting examples suggest variability in the response to information-only CPP notifications, but leave us guessing at the reasons for these differences. Some commentators speculate that the Californian demand response was saturated after several years of conservation campaigns amidst an energy crisis, resulting in relatively little additional change from households receiving an information-only CPP notification (Faruqui & George 2005). However, these and other questions remain unanswered amidst the dominant focus on price responsiveness. Information-only trials, which are characterised by their *absence* of price signals, are only included in trials to 'cross-check' the effect of the price signals (CRA 2005). With the focus firmly situated on price signals, or their absence, there has been little scope to explore other ways of understanding the demand response of dynamic pricing, such as the meanings conveyed by the notification of a peak event itself (Strengers 2010). In order to explore some of these possibilities, we first need to reposition price as a conveyer and distributor of meaning, rather than as a unit of economic analysis in householder decision-making processes.

The meaning of price

In neoclassical economics, prices are 'the outcome of the impersonal forces of supply and demand, which are given to economic actors in a situation of perfect competition' (Velthuis 2004: 372). Consumers respond to prices by weighing up the costs and benefits a product affords (in this case electricity or the services it provides). More recently, the sub-discipline of behavioural economics has gained prominence and popularity (through books such as *Nudge* (Thaler & Sunstein 2008)) in explaining why people don't always behave as the rational model proposes. Behavioural economics is particularly popular because it avoids fundamentally challenging the principles of neoclassical economics, particularly that individuals act to maximise their own utility; markets are the most efficient means of allocation; and markets generate equilibrium as they pursue efficiency (Lutzenhiser 2009). As Lutzenhiser (2009) notes in his critique of behavioural economics, this sub-discipline is interested in 'amending' these basic premises, by focusing on 'correcting' utility maximisation where it is steered off-course by psychological variables such as consumer perception, judgement and choice. These understandings of pricing form some of the fundamental building blocks of the Smart Utopia and of the ideal consumer, Resource Man (see Chapters 2 and 3).

In the discipline of psychology, also essential to the Smart Utopia and its conceptualisation of Resource Man, researchers have tried to shed light on the 'blind spots' of economic understandings of energy use, emphasising the cognitive processes that affect how the information embodied in price enters a consumer's awareness and affects action (Stern 1986). In contrast, sociologists and anthropologists emphasise the social structures, symbolic qualities and cultural dynamics of energy consumption that contravene theories of rational choice and cognitive understandings of price responsiveness (Hackett & Lutzenhiser 1991; Lutzenhiser 1997; Wilhite *et al.* 2000; Wilk & Wilhite 1985).

Building on the ontology of everyday practice outlined in Chapter 4, there is another possible explanation for why and how price 'works'. Price can also be thought of as the conveyer of meaning about energy and indeed about a range of practices-that-use-energy, resulting in new ways of 'handling' (Reckwitz 2002a) (or not handling) this material in practice. This is not a new idea. Prices have long been thought of as conveyers of meaning or as vehicles of communication. However, these meanings and systems of communication are often limited to economic meanings. In neoclassical economics, prices make information available to buyers and sellers, conveying meanings of value and utility. Dynamic pricing, for example, carries meanings about the value of electricity during particular times of day.

Thinking outside economic meanings of price, Velthuis' (2004) analysis of the art sector suggests that prices can be read as a 'text' which refers not only to profit opportunities and scarcity situations, but also to the *qualities* of objects and the people who create and desire them. In Velthuis' (2004) account, prices need to be interpreted to give them meaning, much like language or other symbolic systems which convey social and cultural values as well as economic ones. Applying this understanding to dynamic prices, we might turn our attention to the different qualities of energy that are conveyed through price and how these are interpreted in practice.

For example, price spikes or price 'events' generated by CPP potentially re-materialise the largely invisible electricity system, bringing its vulnerability to the fore. Energy-as-material is afforded new qualities of scarcity and value (Strengers & Maller 2012), which in turn intersect with the other elements of practices-that-use-electricity, reconfiguring what makes sense for people to do. In this way, price can reshape the meaning of electricity as something of worth or value, which in turn intersects with specific practices, which are repositioned as wasteful or unnecessary during CPP events. It might suddenly make sense to turn off the air-conditioner or lights (an unnecessary 'expense' or 'waste'), but not to stop cooking dinner (if this is deemed an essential or immovable activity). Price can thereby result in energy mattering to practice, requiring new (or resurrecting old) skills on how to perform practices differently. For example, in order to *not* participate in air-conditioned cooling practices, householders need other practical skills to perform ways of staying cool that involve minimal or no electricity (Strengers & Maller 2011).

This is not just a case of householders *individually* interpreting price, but rather the active integration of a price signal, or rather the meanings it conveys, into existing and socially shared configurations of practice. Importantly, this is a dynamic process; it is always dependent and contingent on the materials, meanings and skills in circulation and how these are already reproduced by householders in and through their everyday practices (Shove *et al.* 2012). Building on this understanding of price as the conveyer of meaning in an array of practices, the remainder of this chapter considers how and why CPP reveals the elasticity of everyday life, where other disruptions, such as blackouts, have not always been so successful (Trentmann 2009). Why does CPP reveal householders' ability to innovate and adapt, when energy feedback only revealed a relatively limited number of energy-saving actions that obscured the negotiability of other practices (see Chapter 5)? More specifically, why does it place air-conditioned cooling in a negotiable space at some of the hottest times of the year? In short, *why does it work* in seemingly irrational ways?

The exceptional circumstance of CPP

CPP can be thought of as a temporary disruption of 'normal' electricity service (although in some countries it is disruption that is the norm[3]). In this way, it bears many similarities to the blackouts, breakdowns or restrictions that have long been part of the electricity system, and that are a normal feature of everyday life. Leaving aside storms and hurricanes, the average American goes without power for 214 minutes every nine months (Trentmann 2009: 68). These are not always caused by 'technical' problems: squirrels, for example, are responsible for more than 100 local blackouts each year in the US, by gnawing on cables (Nye 2010: 29). Local blackouts are frequent in India, where 50 per cent of the population still live without power and those with it are being rapidly recruited into new practices that demand it (Nye 2010: 225). Heatwaves in Europe are also straining the grid as more and more households air-condition their homes, resulting in several major blackouts during the past decade. This situation is likely to continue globally as (peak) demand for electricity grows, electricity infrastructure ages, and the electricity grid becomes more interconnected and complex.

Importantly, blackouts and breakdowns are not always considered 'disasters'. As historians Nye (2010) and Trentmann (2009) have argued, blackouts can engender outpourings of connectedness and 'mirth-making', as well as disorder, chaos and degeneration. When the lights went out in New York during 1965 many residents experienced 'unexpected elation' and a 'contagion of joy' (Nye 2010: 90, 1). Just over a decade later, in the New York blackout of 1977, there was widespread confusion, fear and discontent, as the absence of light and electrically powered security systems revealed an inherently unstable and electricity-dependent society (Nye 2010). The mood was much more sombre in the epidemic of rolling blackouts experienced by Californians during the energy 'crisis' of the 2000s: society 'did not freeze or paralyze', and there were outpourings neither of joy nor criminality (Nye 2010: 138). However, one survey conducted post-crisis found that 17 per cent of respondents believed the experience had 'possibly improved' their quality of life (Lutzenhiser *et al.* 2002: 8.164–5). These examples show that electricity breakdowns and disruptions are neither universally 'good' nor universally 'bad', but rather have engendered a complex range of reactions and responses at different points in history and in different places.

CPP is distinct from blackouts in a number of ways: first, it seeks to *deliberately* disrupt electricity services, during *predictable* periods of time. It aims to create *ordered disorder*, or *normal disruption*, generating an exceptional circumstance, during which electricity's role in practice is *temporarily* repositioned. To understand this further, we first need to consider the ways in which electricity is normally positioned in relation to household practice in affluent societies. Here, electricity supply continues to follow the 'predict and provide' paradigm that has dominated the electricity supply system throughout the 20th century (Guy & Marvin 1996), framing this resource as an infinitely producible material (Kurz *et al.* 2005). The privatisation of the electricity sector has further cemented this material's position as a commodity that providers are responsible for providing *on demand*. Importantly, these meanings of electricity undermine its position as something that all who use it are responsible for, that is held in common, or that has inherent common value. Thus, when the grid breaks down, householders may blame those deemed responsible for managing it, and resist modifying their own practices to avoid power failure (Nye 2010; Trentmann 2009).

Similarly, the inability of electricity to be produced whenever people demand it, such as during periods of peak demand, can be positioned as a failure of those who provide it. This goes some way to explaining why CPP has been politically unpopular and has rarely gone beyond the trial stage; the implication is that electricity should *always* be available at an affordable cost. Indeed, despite high satisfaction from trial participants and impressive demand response, Faruqui and Palmer (2011: 17) warn that 'a negative mythology has taken root' which has 'prevented dynamic pricing from germinating'. This cautiousness is curious given the proliferation of dynamic pricing for other services, such as for hotels, airlines, rental cars, railways, mobile phones, toll roads, parking meters and even sporting events (Faruqui & Palmer 2011). However, it is understandable in the predict-provide paradigm, where breakdowns, blackouts and temporary disruptions are positioned as endemic systemic failures, and where the rhetoric of supply security maintains the position that electricity *should be* abundant and available at all times (even if it is not) (Strengers & Maller 2012).

Despite this continuing commitment to supply security, electricity systems are also framed within a larger system of normal disorder, where the illusion of secure and unwavering supply is continuously exposed (Graham & Thrift 2007; Star 1999; Van Vliet *et al.* 2005). As Graham and Thrift (2007) argue, the city is in a continuous process of disrepair and repair, where increasingly complex ICT, transport and electricity systems are constantly 'broken', maintained and fixed. Their analysis focuses attention on the everyday experiences of innovation and adaptation which are now normal features of many practices (such as those involving IT). When the computer crashes, an appliance stops working, the hot water system 'blows up', or the lights go out, as is common in both affluent cities and the 'giant system of repair and improvisation' that characterises the Global South (Graham & Thrift 2007: 11), the ubiquitous processes and practices of making, remaking and 'making do' are revealed. Furthermore, meanings of electricity are renegotiated when the infrastructures required to supply it cease to operate *as normal*. Understandings of energy as an abundant, infinite, stable 'thing' shift, uncovering a sometimes scarce, restricted, unstable, and finite substance (Strengers & Maller 2012). In other words, the practices these infrastructures enable are disrupted from their taken-for-granted status when the material on which they depend becomes temporarily unavailable or interrupted (Chappells & Shove 2004b).

The apparent success of CPP, combined with electricity companies' continuing reluctance to implement this strategy beyond the trial stage, can therefore be understood by recognising its unique position between two apparently contradictory meanings of simultaneously maintaining normal disorder, and secure and unwavering supply. Like other temporary measures of 'saving electricity in a hurry' (IEA 2005) or other breakdowns that disrupt the master narrative of 'always-on' infrastructure, CPP does not fundamentally challenge the dominant ethos of meeting, growing and expanding consumer 'needs', but it does place those needs in a negotiable and contestable space for a short period of time. In this sense, CPP balances meanings of electricity as temporarily scarce and valuable, and as a producible, available and abundant material. Energy utilities' reluctance to establish it as the default or standard electricity pricing tariff is not a reflection on its effectiveness, but rather the outcome of these two coexisting meanings, in which regular CPP disruptions potentially challenge the smart utopian promise of maintaining and delivering abundant electricity *on demand*.

Importantly, it is not only the substantial price increase that gives CPP the status of an exceptional circumstance. Another critical feature of this exceptionality is the notification process for an upcoming peak event. As previously outlined, households are generally notified up to a day in advance of a CPP event via one or more ICTs (phone, text message, email, or IHD). In my research with householders participating in a CPP trial in the Australian state of New South Wales, this notification contributed to a sense of emergency and urgency which was likened to a 'blackout', 'power failure' or 'deadly virus' (Strengers 2010: 7319). Despite the use of these terms, these reactions were not reported negatively by householders; indeed, the wider trial from which my qualitative sample was drawn found that 85 per cent of participants' expectations were either met or exceeded (Strengers 2010). This result is similar to that of international trials. Further illustrating the importance of this 'notification effect', the information-only group of EnergyAustralia's Strategic Pricing Study achieved an 11–13 per cent peak reduction during CPP events without any change in price (Strengers 2010). Where notification is combined with price, as it is in CPP, we might then infer that it enhances the position of electricity as a scarce resource during specific periods (peak events).

Similarly, IHDs such as the Energy Orb or EcoMeter (Faruqui & Palmer 2011: 19; Strengers 2010), which change colour according to the price of electricity, become an additional method of communicating and positioning the peak period as an exceptional circumstance, reconfiguring the meanings of electricity for a short period of time and increasing the demand response by a few percentage points. This is one way in which energy feedback and CPP might start to work in tandem, positioning energy as some*thing* that matters to practice.

International research supports this claim. For example, the PowerCentsDC™ Program in the district of Columbia, US, found that 28 per cent of households turned off nearly all of the electricity-consuming appliances during a CPP event – effectively instigating a household-level blackout (eMeter 2010: 67), rather than focusing on the small suite of energy-saving actions as discussed in Chapter 5. Other studies have found that householders report turning off a number of appliances during a CPP period, such as hot water, lights, air-conditioner, heating, fridge/freezer, TV, DVD, stereo, the oven, washing machine, clothes dryer or dishwasher (see, for example, CountryEnergy 2005; eMeter 2010). These responses are similar to other exceptional circumstances, such as those generated during the Californian energy crisis, where there was a 'markedly mixed picture of conservation action' (Lutzenhiser *et al.* 2002: 8–161). Lutzenhiser *et al.* (2002: 8.164) note the 'remarkable resilience and willingness to make changes' among Californians during the crisis, particularly among householders, who were responsible for a significant share of the conservation.

This malleability of practices-that-use-energy in response to CPP is only strange if we assume that the coordination of daily practices is a mostly fixed and enduring assemblage that can only be budged through significant cost-benefit payoffs. However, there is considerable evidence to suggest that exceptionality is a normal part of everyday life. As sociologists and anthropologists have long pointed out, many household practices are carried out with reference to a range of 'special circumstances' that regularly occur in the course of everyday life (Hackett & Lutzenhiser 1985). For example, in Australia, the use of some appliances, such as clothes dryers, appears to be heavily oriented towards exceptional situations, where they are used in 'unusual circumstances', 'emergencies', 'once in a blue moon', or when it's 'rain, rain, rain, rain, rain' (Strengers 2009: 102; participant quotes). Weather and social situations (such as needing a shirt or a school uniform in a hurry) create readily recognisable exceptional events that disrupt normal routines of clothes drying, and render the clothes dryer necessary. The air-conditioner can be similarly oriented around exceptional weather events. As some Australian participants in my research noted, it is there for 'extremes' and for those 'really stinking hot day[s]' (Strengers 2009: 84).

Similarly, Wilhite and Lutzenhiser's (1999) research on 'social loading' reveals a range of exceptional circumstances generated by social events, or social peak loads, whereby appliances acquired and kept on standby for 'social situations such as entertaining, gastronomic extravaganzas [and] erotic encounters' create peaks in household demand (Wilhite & Lutzenhiser 1999: 283–84). The second fridge or the extra freezer for food and beverages needed in case guests pop around are good examples of these special circumstances (although in this case the extra fridge is often on all the time 'in case' of a social visit). Similarly, cooling and heating rooms for guests creates a social peak load in a growing number of countries around the world (Agbemabiese *et al*. 1996; Wilhite *et al*. 1996). In time, these 'exceptional' circumstances may become normal and necessary; for example, the air-conditioner or extra fridge might be deemed 'essential' or used all the time.

While social loads do not always correlate directly with peak loads, the point I wish to make is that social and environmental (weather-related) circumstances routinely interfere with 'normal' practice (and define what 'normal' is), albeit in ways that have tended to increase, rather than decrease, electricity consumption. In this context, the exceptional circumstance of CPP appears as one among many in everyday life. In the following section, I delve further into how practices change during these exceptional periods, focusing on which routines can be shifted to other times of the day or from their normal and seemingly necessary positions.

Shifting routines

Within the Smart Utopia, CPP is often imagined in the context of an increasingly inflexible and electricity-dependent society built on consumer 'needs' and ever-increasing expectations. As the Australian Energy Market Commission explains, 'generation and network assets are deployed to meet the peak demand in accordance with reliability and service standards *desired by consumers* and determined by regulators and governments' (AEMC 2011: 10, emphasis added). To some extent, householders are performing this reality. They have become increasingly intertwined with and dependent on electricity, to the extent that removing it can cause serious disruption and dissatisfaction (Nye 2010). However, electricity (and the ICTs it enables) has also opened up new ways of coordinating, scheduling and shifting practices that were previously unshakable. The rigidity of the 'laundry day' and 'bathing day' have lost their currency in affluent nations (Southerton 2007). Programmable washing machines, dishwashers and clothes dryers can be set to come on at any time of the day and are often interjected between and among other household routines. Freezers, BBQs, microwaves and fridges have opened up the possibility of scheduling and coordinating meals (Hand & Shove 2007). Battery technology enables practices to be performed where the supply of electricity is disrupted or intermittent. And the proliferation of ICTs has brought with it a range of new time-scheduling devices and features, such as mobile phones, synchronised diaries, and programmable appliances (Røpke *et al.* 2010; Southerton 2009). Practices are being shifted in all directions.

However, there is also a counterargument: that modern practices are becoming harder to schedule and coordinate as life becomes increasingly harried and hectic. Southerton's (2003, 2006; 2007; 2009) analyses of the changing rhythms of everyday life challenges this position. He argues that the so-called time squeeze has emerged not as a result of people having less time (indeed statistics show that we work less and have more leisure time than ever before), but rather as the result of anticipating and negotiating 'hot spots' of temporal order around institutionally timed events (such meal or school times), in order to generate corresponding 'cold spots' where 'quality time', 'potter time', 'chill time' and 'bonding time' are experienced (Southerton 2003: 19). In a comparison of the collective coordination of practices in the UK in 1937 and 2000, Southerton (2009: 53) notes the increasing ability of – and pressure on – householders to personally coordinate their practices in 2000, which they describe as a 'roller-coaster ride with moments of harriedness and calm'. In contrast to the 'day in the life of' diaries of householders from 1937, for whom practices were constituted around 'fixed temporal constraints of institutionally timed events and the material hardware of daily life', the 2000s are characterised by the 'de-institutionalisation of many times' and the existence of multiple and overlapping routines which are constantly being made and remade (Southerton 2009: 56, 62). Notably, women feature in this coordinating role, as they bear most of the 'dual burden' of domestic and employment duties (Southerton 2007) – once again challenging the central role of Resource Man, who is largely absent from this literature.

If Southerton's analysis can be more widely applied, it suggests that life in the late twentieth and early twenty-first centuries is characterised by the *increasing* negotiation and negotiability of routines. Rather than generating fixtures and dependencies, electricity has enabled the escalating flexibility and temporality of everyday practices. There are many examples to support this claim. The freezer-microwave combination 're-sequences the temporalities of the practice of meal provisioning', while voicemail, recording devices and automatic timers on appliances have become essential features for shifting practices to other times of the day (Hand & Shove 2007; Southerton 2009: 57). The practice of daily showering, which was virtually non-existent less than a century ago, is now associated with 'speed, immediacy and convenience'; the weekly routine of a Sunday bath has been replaced with 'fragmented moments of washing' (Hand *et al.* 2005: 4). Similarly, the smart mobile phone is generating fluidity and adaptability in the scheduling of meetings and interactions with friends, family members and colleagues (Ling & Yttri 2002). Modern systems of electricity (and water) have allowed for much of this reordering, reorientation and coordination of practice.

A CPP event thus opens up the possibility of generating a 'cold spot' where the normal harriedness of everyday life calms down, and many, if not all practices, are temporally suspended in favour of 'family time' or 'quality time' (Southerton 2003). Although not explicitly investigated, there was some support for this suggestion in my qualitative study of CPP, where households reported using a peak pricing event as an opportunity to take the family out for dinner, light some candles, 'have a bit of fun' or play games (Strengers 2010: 7319; participant quote). Resonating with the social intimacy experienced during New York's 1965 blackouts (Nye 2010), disruptions like CPP can create an opportunity to slow down and reconnect. In this way they reveal the malleability and negotiability of everyday routines; energy feedback, on the other hand, may mask and obscure it (see Chapter 5). The practice of household cooling provides the most pertinent and seemingly irrational example of this negotiability in relation to CPP.

The curious case of air-conditioned cooling

In diverse countries such as Norway (Wilhite *et al.* 1996), the UK (Shove 2003), Thailand (Agbemabiese *et al.* 1996), India (Wilhite 2008a), Australia (Strengers & Maller 2011), and the US (Ackermann 2002; Cooper 1998), there is a similar narrative of the increasing affordability, availability and advancement of the air-conditioner, together with changing housing designs and construction methods, images of modernity, international standards of comfort, and a corresponding decline of competencies and routines about staying cool (Shove 2003). This narrative has played out (and is still playing out) over differing timescales during the late twentieth and early twenty-first centuries. In the US, air-conditioning is now entrenched in all facets of everyday life, constituting a form of 'addiction' in the eyes of some researchers (Prins 1992). In contrast, it has only recently begun a rapid advance in India and China (Wilhite 2008b). A common conclusion is that the entrenchment of air-conditioning into practices of cooling is unstoppable. However, like broad-brush statistics, this global narrative masks significant diversity and divergence in household cooling practices, even in countries like the US, where air-conditioning penetration is very high.

In relation to dynamic pricing, this narrative masks a particularly curious phenomenon. Air-conditioning is often the largest contributor to residential peak load (with the exception of heating in some countries), yet trials conducted in countries where it is regularly used on hot days indicate that it is one of the most negotiable and discretionary appliances during CPP events (along with turning off the lights) (Faruqui *et al.* 2009a; Herter *et al.* 2007; Strengers 2010). Indeed, trials consistently indicate that the demand response increases with higher temperatures and air-conditioning penetration, because householders turn this appliance off (Faruqui *et al.* 2009a; Strengers 2010). Contrary to popular industry opinion, CPP has been shown to sustain reductions in peak usage over time and during prolonged peaky periods (such as a heatwave) (Faruqui & Palmer 2011). Furthermore, research by the Brattle Group in the

US actually cites trials where customers not only maintain their response to CPP, but *increase* it over several years, leading to further reductions in peak usage (Faruqui & Palmer 2011: 20).

To put this another way, on the hottest days of the year, which are arguably the days when householders 'need' air-conditioning the most, householders demonstrate remarkable flexibility in turning up the thermostat, adopting alternative cooling strategies, or simply turning their air-conditioner off. This is not a unique finding. In California during the 2001 electricity crisis, when the air-conditioner was already firmly entrenched in practices of household cooling, 40 per cent of surveyed households either turned it off or used it more sparingly (Lutzenhiser *et al.* 2002: 8.158). In contrast, raising the thermostat was only reported by about 4 per cent of households, despite this action being promoted through pro-conservation advertising. Instead, Lutzenhiser *et al.* (2002: 8.164) suggest that households opted for 'alternative cooling' or even 'rethought' cooling in response to a range of conservation programmes initiated to alleviate the crisis. They suggest that these changes constitute 'actions that conventional energy policy wisdom would expect consumers to be quite unwilling to even consider on the grounds of comfort and convenience' (Lutzenhiser *et al.* 2002: 8.164).

For many utility providers, this contradicts and confounds their rational understandings of consumers and CPP. As one Australian utility provider comments: 'So far it's been almost irrational, if you assume that customers value their air-conditioning use the most on extreme hot summer days. (Strengers 2009: 198; quote from annoymous electricity distributor and retailer)'. There are several explanations for this apparently irrational negotiability of cooling. First, and perhaps most obviously, the link between hot temperatures and air-conditioning is readily made by households in response to CPP, as the participants from my own research clearly demonstrated: 'it's mostly if the weather is very hot or very cold, so it obviously refers to the air-conditioning and heating' (Strengers 2010: 7320; participant quote). Further, householders often understand that heating and cooling are major electricity expenses. Therefore, unlike energy feedback, a CPP event or peak price signal can position energy as something that matters to practices of cooling (and heating), reconfiguring the meanings, materials and skills associated with cooling bodies and homes for a limited period of time, even when understandings of peak demand might be quite poor.

Second, given the rapid advance of air-conditioning into the home, and the ongoing occurrence of blackouts and brownouts, many householders have memories and practical skills of past (and new) practices of staying cool (Maller & Strengers 2013). In this way, air-conditioning can be understood as an already highly provisional and adaptable practice. There are many available materials, for keeping cool still in circulation – particularly windows, shading, the house itself, fans, clothing, cool drinks, cold water – as well as many established practices that keep people cool, such as going shopping, watching a movie at the cinema, swimming in a pool, or taking a cold shower. In my Australian study of CPP, for example, householders engaged in a range of practices to stay cool when a CPP event was called, ranging from going to the beach to placing a wet sheet in front of a pedestal fan (Strengers & Maller 2011). Not only were cooling practices called into question, but householders had the necessary skills, materials and meanings to adapt and perform alternative ways of keeping cool.

Third, the meanings of air-conditioned cooling are arguably still hotly contested in the domestic environment, even where penetration is high. In Australia's southern states, where approximately 73 per cent of households use some form of mechanical cooling (ABS 2011), participants of a CPP trial described domestic air-conditioning as a 'necessary evil', being variously considered unhealthy, noisy, unpleasant, unnatural or something only to be used in 'extremes', as well as a modern necessity (Strengers & Maller 2011). In some situations, meanings of this appliance have taken on strong religious, anti-utopian or political overtones, evidenced by householders referring to themselves as an 'anti-air-conditioner person', or as 'not big believers [in] air-conditioning' (Strengers & Maller 2011: 161). This suggests that, unlike the washing machine or the shower, the use or frequency of which is currently more likely to be contested than its fundamental position in the home (Kaufmann 1998; Shove 2003), practices of cooling may have a more protracted and contested set of meanings which CPP brings to the fore. Like the visible practices of watering the garden or hosing down the driveway, which are deemed wasteful and unnecessary during periods of drought (Head 2008; Taylor et al. 2009), CPP can bring the visible (and often noisy) practice of air-conditioned cooling into a contestable space, rendering it wasteful, discretionary and adaptable, albeit for a short period of time.

A fourth explanation for the negotiability of cooling during CPP events comes from a body of work on 'adaptive comfort', which finds that people can tolerate (and enjoy) a much larger range of temperatures than physiological models of thermal comfort recommend (de Dear & Brager 2002; Humphreys & Nicol 1998; Nicol & Roaf 2007). As de Dear and Brager (2002: 550) argue:

> [P]eople who live or work in naturally ventilated buildings where they are able to open windows become used to thermal diversity that reflects local patterns of daily and seasonal climate variability.

Chappells and Shove's (2005: 33) review of comfort studies supports this contention, finding that people are comfortable at temperatures ranging from 6 to 30°C. Similarly, Cox (2010: 33) reminds his readers that 'homo sapiens is a tropical species'. While intense heat is by no means enjoyable, Cox (2010: 33) points out that it is only in 'extraordinary circumstances' that refrigerated air is required for survival, and even then, it is 'usually under unnatural circumstances of society's own making' that heat kills. Heschong (1979) makes a different case for thermal variation, suggesting that climate-controlled buildings lead to 'thermal monotony', while natural variations create 'thermal delight'. In my own qualitative research with participants of an Australian CPP trial, switching off the air-conditioner for a short period of time was not considered a great sacrifice or source of unbearable discomfort for most householders. As one participant put it: 'You can always go without. I can't ever recall being uncomfortable' (Strengers 2010: 1317). These different perspectives further throw into contention the idea that air-conditioned cooling is a necessary or non-negotiable practice.

It is also important to note that practices of cooling are rarely, if ever, conducted in isolation. Rather, they occur in loosely connected bundles or tightly knit complexes (Shove et al. 2012). Thus, a fifth point is that while some people might turn on their air-conditioner so they can sit and bask in refrigerated air, a more common scenario depicts the air-conditioner as enabling a wider and more physically demanding bundle of practices. The air-conditioner offsets the body (and appliance) heat generated by engaging in a range of household routines, such as cleaning, cooking or working. It enables routines to continue *as normal* without interference from the weather, thereby eradicating the need for the afternoon siesta in hot climates where practices were once seasonally arranged (Shove 2003). It follows then that turning the air-conditioner off during a CPP event is implicated in the broader rearrangement of activities during a peak period, and engagement in different ones. This could involve rescheduling a 'hot spot' (cooking dinner before the peak rather than during it), moving practices to another location (cooking with a gas BBQ), rearranging routines (making a cold salad), or creating a 'cold spot' instead (going out for dinner with the family). While these shifts raise other consumption issues (such as driving to an air-conditioned restaurant), they help reveal the potential elasticity of routines during CPP, and the cooling practices which often enable them.

As a final word of caution, it would be unwise to suggest that everyone can or will change or cease using their air-conditioner on hot summer days when CPP events are likely to be called. Indeed, the demand response during CPP events varies greatly among householders, with one residential study finding that 80 per cent of the demand response is achieved by 30 per cent of participating customers (Faruqui & Sergici 2009 cited in Faruqui *et al.* 2010). Nonetheless, it is clear that far from being a fixed or stable expectation, air-conditioning's role in keeping cool is contested and contestable in the context of CPP and other electricity disruptions. In many ways it is a technology of 'extremes' – something we both love and hate – and something which can simultaneously be deemed necessary and discretionary in extremes of weather and peak demand. Far from being a fixed and immovable practice, cooling is highly movable and malleable.

The considerable diversity and difference in cooling arrangements represents a source of adaptive capacity that policy-makers and utility providers can draw on and seek to enable, rather than a stable and unmoving bedrock for which they must provide. Thinking about the different meanings CPP conveys about energy, through price signals and notifications, as well as how these are integrated into a constantly changing mix of routines and expectations, provides opportunities for utility providers to encourage adaptation and innovation in practice. These insights also open up a new agenda for research and strategy in which the focus is not (only) on price signals, but rather on the malleability of household routines, and on repositioning the meanings associated with energy and its role in practices during exceptional circumstances.

Conclusion

This chapter has departed from conventional economic analyses of price that underpin the smart ontology and view pricing as part of the package of tools and signals that Resource Man requires in order to make cost-benefit decisions about his energy consumption. Instead, the emphasis has been on understanding how dynamic pricing schemes, particularly CPP, intersect with the routines and rhythms of everyday life.

In conclusion, it is worth re-emphasising several points. First, I have argued that CPP temporarily disrupts the meanings of electricity within the practices that use it, attributing to it provisional meanings of frugality and finiteness without challenging broader meanings of abundance and availability. This helps explain why CPP internationally achieves a significant and sustained demand reduction and high consumer satisfaction, while doing little to reconfigure or challenge broader meanings of energy as a non-negotiable need outside a CPP event. The elasticity of everyday life revealed through the temporary disruption of CPP gives energy new meanings of value and scarcity during a specific period of time. When normal service resumes, 'the dominant logic remains one of reactive and incremental expansion, reinforcement and interconnection' (Chappells & Shove 2004b: 140). This remains both an opportunity and a challenge in achieving the aims of the Smart Utopia.

Second, rather than being an unacceptable burden on households, I have suggested that dynamic pricing is interpreted within an already imperfect and changing world, where practices of maintaining and repairing infrastructures and technologies are an (invisible) feature of lived normality. CPP sits alongside a range of other exceptional circumstances, where 'normal' routines are disrupted: when guests come over; when the weather turns wet, cold or hot; when a household member needs their pants dry in time for work or school. Added to this, routines have never been more shiftable with the deterioration of institutionally timed events, a plethora of time-shifting and coordinating devices, and the skills householders already possess to generate 'hot spots' and 'cold spots' of demand. This account challenges narratives that depict CPP as a burden that interferes with non-negotiable needs and temporally fixed routines. Instead, CPP enters a world of already imperfect infrastructure, continual adaptation and innovation, and mobile and movable routines, and it is for these reasons that CPP 'works'.

A final point: redefining electricity as wasteful or unnecessary during CPP periods is strongly and directly connected to practices of household cooling. This is partly because of the timing of these events – on very hot days when householders are more likely to be using cooling devices – but it is also because CPP is interpreted within existing understandings of practices as wasteful, discretionary and malleable. Importantly, this is a dynamic process that may change over time, as other practices shift in their negotiability and perceived need. Cooling practices may not always be so closely connected to the CPP demand response, although the observation that they are now represents another significant opportunity for demand managers.

Viewed in these ways, the 'success' of dynamic pricing schemes depends on their ability to temporarily reposition and disrupt the meanings of energy across a range of practices-that-use-energy, or in relation to one specific practice that uses energy (such as air-conditioned cooling). This process is highly contingent on the practices already in circulation, and on householders' ability to vary those practices. In the following chapter I shift focus again, to the strategy of home automation. Here I find a number of other ways in which smart energy technologies are interfering in everyday routines.

7
Home Automation

Like energy, home automation technologies have an ambiguous status in everyday practice. They are not, in themselves, a material technology, but rather a device or capability that attaches itself to other technologies, such as air-conditioners, pool pumps, washing machines and thermostats. They are characterised by their *in*visibility and *im*materiality, where they are intended to passively and silently operate in the background of everyday life. However, automation technologies can have highly visible effects, bringing new meanings, materialities and skills to everyday practices, and enabling their movement in time and space. In this chapter I put energy to one side, and focus on the 'work' intended for these innocuous devices, the visions of control they seek to embody, and the ways in which they are integrated into, or rejected from, everyday practice.

This discussion is normally absent from the Smart Utopia, where home automation technologies are proposed as part of an unproblematic process of technological substitution, where technologies replace humans in carrying out the tasks of scheduling, coordinating and managing everyday activities that consume energy. As a result, understandings of automation technologies' integration into everyday practice are constrained to analyses of their acceptability and adoption by 'users', and associated quantifiable demand responses. What actually happens to these technologies once they're inside the home is rarely the subject of empirical enquiry by energy utilities, aside from surveys that check householders' satisfaction with and willingness to use these devices. In contrast, social studies of technology continually demonstrate that technology has never operated seamlessly in everyday life. Limited studies of home automation technologies in everyday settings reveal similar insights, demonstrating how these devices can enact different visions and meanings of control (passive, active and others) and perform *multiple* realities.

In this chapter I discuss a series of overlapping performative possibilities for home automation technologies. I begin by depicting the smart utopian vision for smart automation and the reality it seeks to perform. In attempting to automate practice through DLC and smart thermostats, this strategy intends to maintain energy's largely invisible and passive position in practice. In contrast to energy feedback, which seeks to assign responsibility for energy-saving activity to households (see Chapter 5), and dynamic pricing, which seeks to encourage rational decisions about energy in response to pricing signals (see Chapter 6), home automation technologies aim to assign these tasks to technologies and electricity utilities. I warn that this strategy may serve to legitimise practices that are automated by positioning energy as inconsequential in the practices that use it.

I continue by discussing a complementary vision emerging from the smart automation marketplace – one which seeks to make everyday practices more enjoyable and pleasurable by reinventing modern ideals of luxurious domestic life to realise Resource Man's smart lifestyle (see Chapter 3). While this vision is intended to make energy consumption more efficient, I warn that it may also realise more electricity-demanding expectations of comfort, cleanliness, convenience, entertainment and security; expectations that undermine the aims of the Smart Utopia.

I contrast these smart utopian visions for home automation technology with several other possibilities emerging from studies of home automation and domestic technologies in everyday settings. Resonating with the increasing negotiability of routines discussed in Chapter 6, I discuss how smart appliances are becoming enrolled in coordinating practice, by enabling householders to take control of their domestic routines. Automation features here as scheduling and reordering devices that keep activities *under control*, and move them to different times of the day.

I turn next to householders who delegate certain practices to home automation technologies, drawing on research with households that have fully automated homes. Making links with the Golem from

Jewish mythology, I argue that in *assigning control* to someone or some*thing* else, automation technologies can 'act back', taking on human-like roles and blurring the lines between who, or what, is in control. Finally I find that the complicated and fallible characteristics of automation technology can also render it *uncontrollable*, encouraging some householders to ignore or override this technology completely, resulting in its absence from everyday practice. In other examples, automation's complexity constitutes this technology as a high-tech do-it-yourself (DIY) practice in its own right.

This analysis reveals that automation is far more than an innocent bystander in everyday practice. Through a dynamic redistribution of control between people and things, these seemingly benign technologies are involved and enrolled in performing multiple everyday realities.

Automating practice

The smart utopian vision for home automation technologies is summed up by the 'probably-not-as-humble-as-I-should-be opinion' of SmartGridNews.com editor Jesse Berst (2012), who argues that 'we're wasting our time trying to make people smart about energy. We should be making our devices smart about energy.' According to Berst, consumers do not want control over their energy demand, they want 'cruise control'. They should not have to monitor their own energy performance through energy feedback; rather, 'they should tell the system (once) how they want it to respond and then let the system do the watching' (Berst 2012). In this assignment of control from people to machine, the need to monitor and manage energy consumption for efficiency and price responsiveness is not in question; rather these tasks are transferred to technology and industry experts. In Berst's (2012) words, it is the 'system do[ing] the watching' rather than householders. Common arguments to support this position are that householders are too busy and have too many demands on their time to monitor and manage their consumption on a day-to-day basis (Hamilton *et al.* 2012).

DLC and smart thermostats most readily embody this vision of control; they generally involve the automation of cooling and heating appliances, particularly air-conditioning. DLC, which involves the remote control of large appliances, has been offered by Californian utilities since the mid-1980s (Herter 2007) and earlier versions, such as 'ripple control' systems, have been (and still are) used in other countries to manage electricity demand. This technology typically involves the attachment of an innocuous-looking load control device to an appliance, such as an air-conditioner compressor. In some instances fans continue running to allow DLC to operate invisibly (ETSA 2007). DLC is generally considered quite successful, achieving impressive average load reductions of 10–36 per cent (Newsham et al. 2011: 6388). Most DLC programmes target residential air-conditioners: they increase the temperature set-point and/or limit the cycling run time of the compressor (Herter 2007; Kempton et al. 1992b; McGowan 2009; Newsham & Bowker 2010). Hot water systems and pool pumps are other common targets of DLC technology. Many programmes allow customers to override an event if they experience discomfort, but some programmes charge customers for these overrides, or reduce their financial benefits of participating in a DLC programme if they override the system (RMI 2006).

Reflecting this same reassignment of control from people to technology, smart thermostats, also known as two-way thermostats, can communicate with and be controlled by utility providers or demand response systems. Smart thermostats are an extension of programmable thermostats, which have been around for over 60 years. In the US these little devices control almost half of all household energy use, which corresponds to around ten per cent of the nation's total (Meier et al. 2010). The critical difference between a programmable thermostat and a smart thermostat is the latter's ability to be controlled remotely (Hamilton et al. 2012). Smart thermostats can also notify customers (and their air-conditioners) of price events and emergencies, and can be programmed by householders (and electricity providers) to automatically lower or raise the temperature during price events (Meier et al. 2010) – the strategy referred to as set-and-forget (Harper-Slaboszewicz et al. 2012). Smart thermostats typically raise the temperature by 2°C or 4°C during a critical peak period (Faruqui et al. 2010). As such, they are often combined with CPP, where, as part of a suite of 'enabling [automation] technologies', they assist in increasing the demand response from 13–20 to 27–44 per cent (Faruqui & Sergici 2010: 193).

Through both DLC and smart thermostats strategies, householders are positioned as end-user programmers, with the focus centred on the 'usability' of these smart devices. Resource Man features here as an efficient household programmer, able and willing to pre-programme and automate many of his household's day-to-day activities, such as when the thermostat, pool pump, dishwasher or washing machine come on and off. He is simultaneously both in control of his everyday practices and assigning control of these to technology (see Chapter 3).

This vision of rational technological control has been critiqued by digital ethnographers, who argue that it requires an 'a priori specificity and rigidity that conflict[s] with a large body of ethnographic research on the organic, opportunistic, and improvisational ways that families construct, maintain, and modify their routines and plans' (Davidoff et al. 2006: 19). Similarly, other researchers studying the use of digital technology in the home note that the functional and goal-centred ideals of smart automation technologies reflect a masculine image of domestic life. For example, men have been found to focus more on functionality, features and 'the inherent properties of the object' (Livingstone 1992: 119), whereas women are more interested in the significance of domestic technology in their lives and in minimising domestic chaos (Livingstone 1992; Logan et al. 1995; Rode et al. 2004). This body of research challenges the view that everyday practices can be automated, either by electricity utilities or a Resource Man who pre-sets his household's appliances to maximise cost-benefits.

Like other strategies of the Smart Utopia, expectations about how to live are explicitly avoided in this vision for home automation. For example, strategies involving DLC and smart thermostats, which often focus specifically on household air-conditioning, typically subscribe to the American Society for Heating Refrigeration and Air-Conditioning Engineers' (ASHRAE 2004) influential Standard 55 on the thermal environment conditions for human occupancy, which specifies a limited temperature range in which people are found to be comfortable in buildings. The aim is to achieve a demand response 'without affecting the service provided by the appliance' (NERA 2008a: 17). The role of automation technologies is imagined as neutral, with demand viewed as something that is relatively fixed, rather than constantly changing. The assumption is as follows: 'if DLC can operate without affecting customers' thermal comfort levels a DLC rollout would unambiguously result in a reduction of greenhouse gas emissions, as it would result in overall reduction of demand rather than a shift' (NERA 2008b: 7). The assumed 'unambiguous' nature of DLC and other automation technologies is what makes them so popular among utilities, who are understandably concerned about unpredictable demand from householders. However, these devices have a much more ambiguous set of possibilities.

One possibility is that, by monitoring and attempting to control electricity demand, meanings of rationality and efficiency are brought into the home and normalised as accepted forms of electrical participation, or the accepted ways of saving energy. 'Managing' demand and participating in energy conservation might become framed around cycling an air-conditioner on and off within a narrow temperature band, once again resonating with Marres' (2011: 527) depiction of 'the change of no change' embodied in material devices such as those that provide energy feedback (see Chapter 5). From this perspective, participation is no longer contained within the limited suite of energy-saving actions one can and should take to reduce energy, as argued in Chapter 5, but is understood in terms of one's *inaction*, or one's assignment of action to another thing or things, specifically DLC and smart thermostats. In the terminology of everyday practice (see Chapter 4), skills around how to stay cool are assigned to some*thing* else: meanings about how to use energy and air-conditioners responsibly during peak events reflect ideals of efficiency that involve the continual use of electricity, while other things used to stay cool, such as windows, eaves, doors, showers, sprinklers, pools, clothing, ice, water and passive solar building designs, are potentially displaced.

Unlike CPP, which can generate an exceptional circumstance in which everyday practices are negotiated during a specific time period (see Chapter 6), air-conditioned cooling practices are not called into question during peak times when DLC or a smart thermostat is in operation. Instead, they are potentially legitimised and further entrenched as normal and necessary practices (Strengers 2008). Thus, while automation technologies are designed to enable householders to continue performing their current practices *as normal*, they may also recalibrate this normality. Herter's (2007) analysis of residential CPP tariffs in California provides some evidence of this. She found that households on a CPP tariff who also had a smart thermostat responded substantially on hot days (when their smart air-conditioning thermostat was likely to be working) but less on cooler days (when it was not). Herter (2007) hypothesised that households with a smart thermostat relied on this device to respond to events on their behalf and therefore did not pay much attention to the CPP signal. As a result, when temperatures were low and air-conditioning was not used, there was little or no demand reduction (Herter *et al.* 2007). In this example, technological efficiency takes precedence as the (only) means of responding to, and participating in, CPP events.

Importantly, this is not a universal conclusion. Other studies find that smart thermostats, when combined with dynamic pricing programmes, substantially increase the demand response (Faruqui & Sergici 2010: 193). It remains unclear exactly how automation technologies reorient participation in dynamic pricing programmes. This is an area of enquiry where there is little available evidence and therefore great scope for further research.

As a final word of caution, there is some evidence to suggest that householders may reject these 'passive' devices in their everyday lives, as well as the meanings of efficiency and rationality they embody. In most cases, DLC and smart thermostats have been positively received; however, one example illustrates how rejection can occur. When California introduced one of the toughest building codes in the country and attempted to introduce a 'programmable communicating thermostat' as part of the provisions, it met with strong resistance. The thermostats, sold through traditional retail channels rather than supplied by the utility, were intended to enable the utility to control load at critical peak times, as well as allow customers to set heating and cooling offsets during price events. However, the plan was scrapped when news of the proposal reached the media, which described it as a 'Big Brother' attempt to ration energy use and take choices away from consumers. The outcry over consumer privacy forced the Californian Energy Commission to scrap the controversial part of the plan: that consumers would not be allowed to override the utility-specific temperature setpoint during an emergency event to avoid a blackout (Meier *et al.* 2010). Similarly, many householders opted out of California's DLC programme or refused to curtail their energy demand when requested during the energy crises in 2001 (Goldman *et al.* 2002).

These examples demonstrate the politically oriented meanings automation technologies potentially bring to everyday practice, where they can be interpreted as methods of monitoring and controlling householders. This resonates with Foucault's (1995) analysis of technology as part of a Panopticon of discipline and control, in which both providers and consumers of electricity are watching and being watched. Despite intending to be passive, smart energy technologies can be thought of as a form of 'governmental technology', intended to keep populations under the control of the state (Rose & Miller 2010: 273). This is a vision that can be resisted by householders, as outlined above, as well as confounded by the other ways in which automation technologies encounter everyday life.

Enhancing practice

In addition to smart automation technologies like DLC, which are intended solely for the purpose of energy demand management, smart energy proponents also emphasise the increasing ubiquity and importance of smart appliances and the fully automated smart home in achieving the aims of the Smart Utopia. Like other smarts, these appliances are characterised by two-way communication, which allows householders, technologies or other parties to control them remotely. In the Smart Utopia, these appliances can communicate with dynamic pricing tariffs, enabling what is referred to as 'prices-to-devices', whereby appliances '"listen" to the price of electricity and operate accordingly' (Hledik 2009: 31).

The Association of Home Appliance Manufacturers (AHAM) adopts a deceptively neutral definition of a smart appliance, describing it as 'a modernisation of the electricity usage system of a home appliance so that it monitors, protects, and automatically adjusts its operation *to the needs of its owner*' (in Hamilton *et al.* 2012: 409–10, emphasis added). This is a utilitarian position that promotes a reality in which 'needs' are solely determined and controlled by individual home appliance owners. In this last part of the definition, the AHAM avoids any recognition of, or responsibility for, the role that appliance manufacturers play in establishing what these needs are. However, in this current depiction of an electrically enabled and technologically mediated lifestyle, energy utilities and appliance manufacturers build on a long history of visions intended to sell more electrical stuff – attempts that have played a profound role in establishing expectations of normality and new 'needs' (Forty 1986; Shove 2003).

More explicitly, the (smart) appliance industry's current marketing vision reinvents the old idea of creating a labour-free home environment, where electricity provided 'the modern housewife with a perfect servant – clean, silent and economical' (Forty 1986: 207). Close to a century after this vision was first promoted to households, home automation company Control4 (2013) rekindles these ideas and adds new ones by inviting its customers to 'imagine a house that remembers to lock itself at 10pm. Shades that close as the sun hits. A home theatre that takes care of lights, sound and picture in one touch.' The company's tagline – 'life is just better with a little more control' – references the desire to assign housework to someone, or something else, featuring automation as the way to be 'in control' of domestic activity (Control4 2013). These ideals are further expanded to encompass not only the control of household labour, but of home security, comfort and entertainment. The outcome or 'selling point' for householders is 'unprecedented levels of convenience', with energy bill savings and increased home security listed as important side benefits (Harper-Slaboszewicz *et al.* 2012: 393).

Implicit in this vision for home automation is an underlying commitment to consumption, where more (convenience, comfort, entertainment, security) is better than less, and in which energy management and efficiency feature as byproducts or side effects of this improved quality of life. Visions of 'normal' and desirable hominess are central, as are meanings of technologically controlled, efficiently delivered and, above all, electrically powered comfort, security and entertainment. In this way, this vision does not attempt to maintain a neutral position on how we should live at all; rather it promises, promotes and reinvigorates utopian ideals of seamlessly integrated, harmonious and labour-free home life, which is efficiently run, silently managed and enables new forms of electrically enabled pleasure and indulgence (Berg 1994; Dourish & Bell 2011).

Importantly, as I suggested in Chapter 2, we should not conclude here that this vision will unfold as intended; it is a possibility, or rather a series of possibilities. Just as utopian narratives of an efficient smart lifestyle or the uncomplicated automation of practice mask diversity and contradictions, so too do simple dystopian narratives of assimilation, addiction and disconnection. Yes, technologies enter everyday life in ways that both perform and confound the visions of those who make and promote them, but this does not mean we can (or should) categorise them as either 'bad' or 'good', or 'successes' or 'failures'. Nonetheless, there are good reasons to be cautious with the vision espoused by automation companies like Control4. For example, Røpke *et al.* (2012; 2010) find that ICTs are enabling 'a new round of household electrification' and giving rise to other forms of electricity consumption, such as the running of server parks and sending masts. While there are no definitive figures on this, Willum (2008: 14) provides the following 'rule of thumb': 'when 1 [kilowatt hour] kWh is consumed in the residence, 1 kWh is consumed to manufacture, transport and dispose of the hardware and ½ kWh is consumed to run the Internet and the applied ICT infrastructure outside the residence.' These are 'hidden' energy costs and impacts of the Smart Utopia that deserve our critical attention.

However, for now I want to consider other possibilities for home automation technologies. This means stepping outside these visions and turning to studies of technology in everyday life, where researchers have explored the dynamic ways in which digital and automation technologies enter the home and the practices performed within it. It is here that we find further possibilities for home automation's role in everyday practice, and other meanings of control in circulation.

Coordinating practices

One of the ways in which automation technologies, particularly smart appliances, are entering the home is as a form of coordinating or mediating technology across a variety of everyday routines. Davidoff *et al.* (2006: 19) argue that coordinating technologies are becoming a necessity in a world increasingly characterised by the feeling of being 'out of control due to the complex and rapidly changing logistics that result from integrating and prioritizing work, school, family, and enrichment activities'. In their study of dual-income families using smart home technologies, these researchers distinguish between the common human category of technology users, who are thought to set clearly defined goals, and families, who resist goals and change them over time. Unlike users, the logistical and organisational challenges of running a home lead families to 'aggressively adopt and experimentally use new communication and coordination technologies', not because they are seeking to manage their energy consumption or improve their comfort, convenience and entertainment, but because they need to *take control* of their everyday lives (Davidoff *et al.* 2006: 22). Like mobile phones (and in some cases enabled by them), home automation technologies feature here as scheduling and organising devices that allow everyday routines to be shifted and adapted. In contrast to the smart utopian vision of automating what are seen to be largely fixed and immovable routines, Davidoff *et al.* (2006) refer to smart technology's role in creating *increasing* degrees of flexibility around sites of activity which blur the boundaries between home and work. These researchers conclude that 'to give families a sense of control over their lives, a smart home system will have to both support the concept of routine, but not bind families to that notion. Such a system will need to allow plans and routines to evolve organically' (Davidoff *et al.* 2006: 28). This system, they argue, will need to allow for 'battles' over the thermostat, substantial innovation, 'constantly shifting targets', failed routines, and many time-intensive tasks that are nonetheless 'vital to our identities as Mums, Dads and Families' (Davidoff *et al.* 2006: 29–31). Unlike the rational and rationalising visions of control central to the smart ontology, control is positioned here as something that is fluid and dynamic, embedded in daily routine, and not explicitly focused on flows of energy in and around the home.

Similarly, Leshed and Sengers' (2011: 912) research reveals the cultural complexities associated with being 'busy', which they describe as a social norm that cuts across many lines of domestic activity. They argue that productivity tools have a broader role than the generation of goal-oriented 'to-do' lists. Rather, they are also used by householders to 'renegotiate their goals and priorities, feel socially committed, manage ever-changing real-life interactions, *feel in control*, and organize not only what they do, but also who they are' (Leshed & Sengers 2011: 912; emphasis added). Similarly, automation technologies might be thought to generate 'downtime' and 'slowness', not necessarily of the luxurious kind espoused by home automation companies, but more practically experienced as a moment to 'charge the batteries' (Leshed & Sengers 2011: 912). In this example, control does not feature as the technological management and surveillance of the home, but as the ability to dynamically schedule time when no management is required, and where everyday life can be freed from the demands of people and technology. This is consistent with Southerton's (2009) analysis of 'cold spots' in activity, and connects with the possibilities discussed in Chapter 6, where CPP events generate temporary disruptions in everyday practice, during which time automation and other ICTs can play a coordinating role.

However, this coordinating role of home automation technologies, while enabling practices to shift to off-peak times of the day, can also increase energy demand overall. For example, Røpke and Christensen (2012: 359) point out that 'ICT contributes to both the increasing complexity of everyday life and to the handling of this complexity – making possible the management of more practices'. Home automation technologies can be thought of in this way, where they feature as devices intended to manage increasingly complex time-space arrangements. A busy parent can be making plans to pick up a child by SMS while cooking dinner. At the same time, automation may be enabling another domestic activity, such as laundering, to be performed in the background. Røpke and Christensen (2012) note that this ability to perform multiple practices at the same time presents new opportunities for energy to be consumed. The concept of multi-tasking and making use of 'dead time' lead these researchers to suggest that 'more energy can be spent per unit of time' (Røpke & Christensen 2012: 359). These potential energy implications deserve further investigation in empirical research.

Delegating practices

A more complicated field of possibilities for home automation technologies can be found in fully automated homes. Communities of Orthodox Jews have been using home automation technology for decades to coordinate the religious practice of resting on the Sabbath, and to maintain the modern interpretation, which is that it is forbidden to turn electrical devices on or off on this day (Woodruff et al. 2007). In these and other examples, householders delegate practices they would have previously performed themselves to these technologies. Unlike the electricity industry's intentions to passively automate practice through DLC and smart thermostats, this is not about participating in electricity management or delegating control of household energy consumption to a 'Big Brother'. Rather, it is about assigning everyday practices to technology – a seemingly banal task that can lead to these technologies 'acting back', making their own demands on practice and embodying other human-like roles.

An example of this can be found in Woodruff et al.'s (2007: 529) qualitative study of home automation technologies in Orthodox Jewish families, where the Sabbath is described by their participants 'as a time of peace, relaxation, and reflection', which resonates with the increasing desire to generate downtime and cold spots of activity discussed above. By pre-scheduling and timing a variety of devices to turn off or on a day in advance, these researchers find that 'Orthodox Jews strive to clear their minds to focus and reflect on larger issues' (Woodruff et al. 2007: 529). Automation technologies allow Jewish households to transcend the mundane demands of technology, or in Woodruff et al.'s (2007: 531) words: 'more technology provide[s] the illusion of less technology'.

The concealment of everyday activities is one of the attractions of automation for this community; however, this technology also reveals and enhances sensory experiences, adding ambience and atmosphere on the Sabbath through the automation of lighting, water fountains and the scheduled production of meals – thereby reinforcing and establishing new meanings (of aesthetics) and routines (of meal time), and reassigning skills (of turning things on and off) to technology. For example, appliances and lighting that would have otherwise remained on (or off) for the entirety of the Sabbath are pre-scheduled to operate in specific rooms at specific times, both saving and using electricity associated with the Sabbath's activities of 'going to synagogue, spending time with family and friends, studying religious materials, reflecting, taking naps, and going for walks' (Woodruff et al. 2007: 529). As well as enabling many practices, this has the unanticipated effect of making new demands on practice.

In making sense of these dynamics we can draw insight from the Golem of Jewish mythology, who is represented as a human-like creature made by Rabbis from clay and 'created by magical art' to silently serve the Jewish people (Scholem 1965: 159; 1966). Initially, this was intended to be a relatively straightforward process of technological substitution, where the Golem carried out tasks previously performed by humans. He could respond to orders and sort them out, but that was the extent of his abilities. But here the story gets interesting. The Golem, who was intended to be controlled by his creator, had 'a dangerous tendency to outgrow that control and develop destructive potentialities' (Scholem 1966: 63). In one version of the story (the Golem of Prague), the Rabbi who had made the Golem had to destroy him to stop him from 'running amok' (Scholem 1966: 63). This was done by tearing a slip of paper with the 'Name of God' on it from the Golem's mouth, an act that rendered the Golem an inanimate body of clay (but when reinserted, brought him back to life) (Scholem 1966: 63).

Automation can be imagined as an extension of the Golem, where it features as a new intelligent household 'slave'. In this case, technology takes over from the human-like Golem, who was given his day of rest on the Sabbath, just as labour-saving devices replaced the shortage of domestic servants in a burgeoning middle class during the early twentieth century (Forty 1986). Like the Golem, automation technologies can be easily turned on and off, rendering them either inanimate or animate in the practices they seek to perform.

However, home automation technologies are not passive slaves: they act back, taking on human-like characteristics and embodying Golem-like agency by guiding, facilitating and even recommending specific actions, despite the fact that they are 'controlled' by their 'master'. An example of this is found in Woodruff *et al.*'s (2007) research, which found that some householders viewed a high-end and highly complex automation system as an extension of the system's designer – Mr Herschel – with whom these householders had a personal relationship. In some of these households, the automation of lighting or other electrical appliances was viewed by participants as comments from Mr Herschel on their behaviour, partly because Mr Herschel was largely responsible for programming the system on their behalf, and on fixing it when it broke down or 'acted up'. One participant even called it 'the Herschel System...Well, the truth is, I think of it as Mr. Herschel himself' (Woodruff *et al.* 2007: 533). Traces of Golem mythology resurface in this example, as technology that is 'produced by the magical power of man' takes on its maker's human shape (and flaws) (Scholem 1966: 63).

Automation technologies can embody other Golem-like roles too. In Woodruff *et al.*'s (2007) study, an automated light turning off in a recreation room sent a 'message' to the children within it to go to bed, thereby performing the role of the parent or carer. Similarly, lights turning off or dimming after the Sabbath meal were an indication that it was time for guests to leave, performing the role of a polite host. These human-like roles were not viewed negatively or in reference to a judgemental Big Brother. Rather, householders associated automation with 'caretaking, anticipation, and guidance – roles such as servant (sometimes quite a wise servant), mother, and wife' and 'occasional allusions to more godlike or omniscient characteristics' (Woodruff *et al.* 2007: 533) such as those espoused in the mythology of the Golem.

Similarly, Mozer's (2005) experience of living in a self-learning smart home assigns the human-like role of a housemate to this technology. The environment Mozer refers to is one where the smart home learns and predicts its occupants' behaviours, operating seamlessly and invisibly in the background. However, it is precisely its invisible workings that make the automated home so visible to Mozer. For example, he describes how he 'found it disconcerting when ACHE [Adaptive Control of Home Environments] would incorrectly predict my passage into another room and lights would turn on or off in an unoccupied area of the house', and notes that 'when ACHE is disabled, the home seems cold and uninviting' (Mozer 2005: 291). In contrast to the idea that this automated system can and should learn to fit in with and support his routines, Mozer (2005: 292) describes how he found himself trying to fit in with the routines 'learnt' by the system:

> For instance, if I were at work at 8 p.m., I would realize that under ordinary circumstances, I might have left several hours earlier; consequently, ACHE would be expecting me, and I felt compelled to return home. I regularized my schedule in order to accommodate ACHE and its actions.

Koskela and Väänänen-Vainio-Mattila (2004: 239) report similar findings in their study of home automation technologies, where the invisibility of automation simultaneous renders itself visible when things automatically turn on and off as if the house has a 'life of its own'.

Like the Golem, technology features here as 'a technical servant of man's needs', embodying human-like roles and performing human-like tasks (Scholem 1966: 64). However, this is an 'uneasy and precarious equilibrium' (Scholem 1966: 64). Technologies, like the Golem, can act back, reflecting our own practices and routines back at us in ways that are potentially transformative.

While we might not go so far as to suggest that automation technologies have the Golem's 'destructive potentialities' that dominate practice or 'run amok' (Scholem 1966: 63), they clearly bring new ideas to existing practices by, for example, allowing the integration of new materials (water fountains and pretty lights) to meditation and relaxation practices. The mythology of the Golem reminds us that these enhancements of practice cannot (only) be attributed to the designers and makers of these technologies or to some in-built 'scripts' (Akrich 1992). Automation technology's role in practice is much more subtle than that. In the examples discussed above, they take up the position of a modern-day diplomatic Golem who is responding to and making suggestions that potentially reconfigure the existing constellation of practice elements. This is not a uni-directional imposition or reassignment of control; it is a cyclic process in which control is circulated and redistributed between people and technology, through practice.

Absence from practice and DIY practice

There are two final possibilities for home automation technologies as they encounter everyday life as complicated, unworkable, uncontrollable or fallible technology. The first is that, because they are overly complex, break down, or do things that that annoy and frustrate householders, these technologies are switched off and regulated to the back of the cupboard – in short, they do not perform anything at all. For these same reasons, the second possibility is that automation technologies come to constitute a DIY practice of repair and innovation.

In the first of these possibilities householders may lack the skills to use and fix automated functions when they break down or 'play up', or they may not be clear on why they would want to use them in the first place. This is a common finding in studies of other computer systems and ICTs, where the capabilities of new technology are often poorly understood by householders (Meier *et al.* 2010), or prone to misinterpretation or failure (Graham & Thrift 2007).

For example, Meier *et al.* (2010: 9) find that 'occupants find thermostats cryptic and baffling to operate because manufacturers often rely on obscure, and sometimes even contradictory, terms, symbols, procedures, and icons'. In their review of programmable thermostat usability studies, these researchers cite an impressive list of thermostat (and energy) misconceptions and complaints that illustrate how householders are not always (or often) expert operators of systems, and often use complex thermostats as an on/off switch. In this situation, a smart thermostat is no different from a 'dumb' thermostat, being so complicated that its features are considered unusable.

This is not a new phenomenon specific to smart automation – Kempton et al. (1992a) made a similar point over two decades ago in their study of air-conditioner operation with New Jersey (US) residents. They found that residents' 'folk physiological theories' about their bodies were far more important in determining how the air-conditioner was used than the built-in thermostat function (Kempton et al. 1992a: 189). From a practice perspective, we might conclude that the meanings embodied in a built-in thermostat or other automated device conflict with those that already exist in practice, such as what type of cool air (refrigerated, dehumidified, fanned, 'fresh') and how much air is needed for health and comfort. Additionally or alternatively, the skills required to operate these devices might be lacking. By making demands too far removed from the practices it seeks to automate; automation remains absent from practice.

A second possibility is that the complicated and/or failure-prone system of automation can constitute its own DIY practice, requiring constant interference and attention and rendering the system highly visible and demanding. Rather than rejecting or ignoring this technology, this may lead some householders to participate in practices of maintenance and repair located specifically around this suite of technologies. For example, Woodruff et al.'s (2007: 530) participants note that the automation system X10 is 'notoriously unreliable' and depends on the competence of 'tech-savvy "do-it-yourself-ers"' who have impressive stories of both success and failure.

Like other DIY home-improvement practices, the 'project' of home automation involves 'sweat, sawdust, frustrations and satisfactions generated through the active combination of bodies, tools, materials and existing structures, all of which are implicated in repairing, maintaining or improving the home' (Shove et al. 2007: 49). More specifically, it requires a very specific set of skills about how to fix automation devices, technological materials including the automation system itself, and meanings about the value of engaging in the practice and how to do it well. Such projects are emergent in the sense that they never go exactly to plan, and because they always involve complexity, exploration and uncertainty. Importantly, DIY automation does not stand alone as a practice, but necessarily intersects with a range of other practices that it seeks to automate. This arrangement is inherently unstable, perhaps more than practices usually are, because DIY automation is deliberately experimental and does not always work, or at least not always as planned (Shove et al. 2007). This is likely to generate a fluid field of possibilities for automation's role in everyday practice.

Performative possibilities

By attempting to operate in the background of everyday life, automation technologies can be unintentionally foregrounded as a material that makes implicit and explicit demands on and in practice. Similar to electricity, automation's stealthy and silent approach positions it as an immaterial material (Pierce & Paulos 2010) in the practices it seeks to automate, making its integration into practice sometimes difficult to pin down. By attaching itself to or merging itself with existing appliances in the home in order to schedule, organise, anticipate or otherwise enable practice, automation aims, often unsuccessfully, to maintain a passive role in practice. Embodying new ideals of control and coordination, home automation technologies confront other materials, meanings and competencies, cut across and navigate around and between many practices, and, like other materials, must be integrated into practices by those who perform them.

It is clear from this chapter that there is no one way in which automation enters the home and the practices performed within it. Rather, these technologies can enact *multiple* realities and perform different visions of control, sometimes simultaneously. A smart washing machine can embody ideals of rational and efficient control of energy demand as well as a better, labour-saving life. It is simultaneously a means for householders to take control of doing the laundry, scheduling it between other domestic activities. Further, householders can delegate control of doing the laundry to someone or something else. The washing machine can also take control, by recommending when it should be used, on what settings, and for how long, thereby bringing new meanings and competencies to

134 *Smart Energy Technologies in Everyday Life*

the practice of laundering. It can also be completely ignored, with householders judging the enhanced automation functions to be uncontrollable, and thereby ignoring them and using the appliance 'as normal'. Further, a smart washing machine's complicated and seemingly erratic workings may constitute it as a DIY project – an activity to be worked on – in which automating the laundry is reconfigured as an ongoing experimental process. Most importantly, automation technology can mean all or some of these things at once.

Despite these performative potentialities for home automation technologies, the Smart Utopia positions automation as a means to promote and perform a technologically oriented and mediated way of life, in which technology solves a range of electricity management problems. This vision not only excludes other potential realities that smart automation technologies might enact, but also attempts to perform or simply maintain a very specific vision of 'the good life'. In doing so, these technologies silently sidestep and sideline debate and engagement about other normal and desirable ways of living, as so many other technologies have done before. In considering how automation technologies can assist in achieving the aims of the Smart Utopia, these are lines of enquiry that require our attention. The following chapter develops another field of possibilities for smart energy technologies in everyday life, focusing on the smart strategy of micro-generation.

8
Micro-generation

The provision of micro-generation, micro-grids, distributed generation or 'local energy' has predominantly been a topic for engineers and technologists rather than social scientists. It has also been the subject of economic and regulatory debate, where opportunities for the trading of micro-generated energy are discussed in similar terms to the ways in which other commodities are traded on stock markets (Ramchurn *et al.* 2012). The 'human side' of these technologies sits outside these issues; it features as either the social acceptance of renewable or small-scale generation technologies (Caird & Roy 2010; Sauter & Watson 2007; Wolsink 2012) or a type of behavioural change resulting from contact with, and awareness of, a localised energy supply system (Keirstead 2007).

This chapter opens up other possibilities for thinking about the role of micro-generation in everyday life, focusing specifically on the delivery of this strategy at the household scale. I outline how different energy systems constitute *energy-making practices* – that is, practices of making energy in different ways. By positioning the energy produced by micro (and macro) energy-making practices as a material 'thing', I seek to develop a more nuanced understanding of the role that micro-generated energies play in everyday practice, where energy-as-material meets with constellations of other materials, meanings and skills (see Chapter 4). Developing this analysis, I find that micro-generation presents another set of possibilities for making energies that matter, and do not matter, to practice. By 'mattering' I mean the ability of energy to become integrated into everyday practices in ways that shift or shed energy demand. I begin by depicting the ways in which this strategy is understood within the smart ontology, where micro-generation is positioned as a means to both activate and pacify the smart energy consumer.

Activating and pacifying the consumer

The term 'micro-generation' is used to describe small-scale systems of electricity provision (generally up to 50 kilowatts) and/or heat (up to 45 kilowatts thermal), which are used by households or community buildings to generate power on a specific site. A micro-grid is slightly different, comprising 'a collection of geographically proximate, electrically connected loads and generators' (Platt *et al.* 2012: 186). The terms encompass a range of potential technologies, such as solar photovoltaic (PV) panels, solar thermal heating, micro-wind power, biomass-fuelled boilers and micro-combined heat and power systems (Bergman & Eyre 2011). These technologies are united by the small scales and sites in and on which they operate. However, they differ markedly in terms of their relationship to the wider electricity grid by being on- or off-grid, the sources of energy they draw on (wind, solar, biomass), and in terms of their physical or technical characteristics. While this chapter draws on research involving a variety of different micro-energy systems, my focus is on the provision of micro-generation at the household scale.

The vision for these technologies is no different from the other smart strategies discussed in this book, except that here the active role intended for consumers exists in tandem with a more passive one. A critical starting point is that the consumption of energy is, like other consumption, an inherently passive process, made possible through the centralised control and supply of electricity. The goal then becomes one of maintaining this passivity, and/or transforming consumers into active producers (House of Commons 2007) or prosumers (Ramchurn *et al.* 2012) through access to micro-generation. The term 'active' refers to the smart ontology's specific focus on enrolling consumers in micro-resource management. Here it means two things: first, it refers to the consumer's active acceptance of micro-generation technologies as opposed to the more passive acceptance of large-scale energy projects; and second, it refers to the smart utopian vision for an active Resource Man who is engaged in managing and maintaining his micro-generation technology (see Chapter 3). This maintenance and management can be mediated by feed-in tariffs that encourage householders to export power at specific times of the day, or energy consumption feedback that informs them when they are producing and consuming energy. Such tactics are framed in information-deficit and rational choice terms, both of which are central to the smart ontology (see Chapter 2). The goal is to realise Resource Man's full potential, with energy production matched with consumption through the use of pricing, information and technology tools. In order to achieve this aim, householders require 'reliable and impartial advice' to 'make the right choice of local energy system for their home' (House of Commons 2007: 31). Once installed, information is also required (and requested) in the form of regular feedback about how much energy is being produced, consumed and/or sold or supplied to the grid (Caird & Roy 2010). Some studies indicate that this feedback is critical to maintain the active 'link' between micro-generation and behaviour and maximise the export potential of this generation into the wider grid (Bahaj & James 2007; Bergman & Eyre 2011).

In contrast, micro-generation can also enhance consumer passivity by providing 'plug and play'[1] solutions to demand management problems which maximise consumers' 'private economic benefits' (Leenheer *et al.* 2011: 1983). Resonating with the technologies and goals of home automation (see Chapter 7), the vision for passive micro-generation involves a self-sustaining and self-managing power system that is linked to other automated devices in the home. For example, Bahaj and James (2007: 2126) cite the need for 'intelligent consumer units [to] take control [of energy production and consumption] on behalf of the consumer' if, for example, householders are not home during the day. Similarly, Platt *et al.* (2012: 205, emphasis added) discuss how micro-grids can 'offer significant benefits to the end-user, *helping to isolate them from utility grid issues*, manage multiple on-site loads and generators, or improve local power quality'. Rather than empowering consumers to take control, this might involve providing utilities with more control over energy systems and appliances in order to 'guarantee desired behavior of the system' (Platt *et al.* 2012: 205).

Micro-generation strategies thus straddle a passive and active vision for the smart energy consumer. Resource Man is both in control of his consumption and production, and assigning control of these activities to someone or something else, much like the role of an energy engineer (see Chapter 3). There is some evidence of householders performing this vision when they have micro-generation installed on their property. For example, researchers report the tendency for some users of these technologies to be 'technophiles' (Dobbyn & Thomas 2005: 8) or 'technology enthusiasts' (House of Commons 2007: 29), who are interested in DIY energy-making and automation. However, others perform an entirely passive role. As studies of energy feedback discovered (see Chapter 5), Abi-Ghanem and Haggett (2010: 16) also find that some householders with micro-generation systems do not have anything to do with these technologies or its associated display monitor (IHD). In other words, in some households micro-generation might not perform anything in particular at all.

A key problem with this active–passive vision is that it runs the risk of narrowing the issue to one of activating or pacifying energy consumers in the development and deployment of micro-generation technologies. It also implies that individual consumers can somehow be activated or pacified, leading to a continual need for the electricity industry to either placate consumers or enrol them in new forms of energy-saving 'effort'. This focus overlooks the routinised ways in which energy is *already* integrated into everyday practices – ways that cannot be explained in relation to either active decision-making processes or complete passivity. Furthermore, this vocabulary sets up a problematic dichotomy (active versus passive) that is counterproductive in attempting to understand the ways in which micro-generation technologies and the energy they produce come to matter, or not matter, in everyday practice. Rather than asking how to engage or disengage consumers, a more promising line of enquiry involves thinking about how energy systems participate in everyday practices.

Some researchers have sought to understand similar dynamics by positioning technologies, infrastructures and other non-human actors as 'intermediaries' between the realms of production and consumption (Guy *et al.* 2011), acting as mediators in issues such as electricity demand management. From this perspective, micro-generation can simultaneously enrol householders in and unburden them from various forms of effort and responsibility in energy management, thus generating both active and passive modes of participation. For Marres (2012a: 106; emphasis in original) the challenge is one of examining 'how material entities *become invested* with specific capacities, like powers of engagement, in particular settings and at certain times'. Applying this thinking to micro-generation involves considering how these systems of power generation become invested in or divested of specific modes of participation. This is fruitful territory.

However, by positioning micro-generation systems as intermediaries between production and consumption or as devices with participatory investments, the focus is likely to remain on the intermediation of, or participation in, energy issues. This will reveal important insights about how householders participate in energy management problems, and what types of relationships and roles different energy systems enable and prioritise for the providers and consumers of power (Marvin *et al.* 2011). However, it tells us little about how householders engage in other forms of 'everyday' participation as a result of having access to micro-generation. It does not tell us what micro-generation systems mean for how householders do the laundry, cool the home or cook dinner.

In order to get from energy production to everyday practice we need a different set of questions and concepts. We need to ask how micro-generation systems become integrated, or not, into domestic routines. One way to answer this question is to think of energy systems as constituting practices of making energy. What then become important are the meanings, skills and materials of energy-making practices, the energies that they produce, and the ways in which these energies are integrated into everyday practice. The following discussion develops these conceptual ideas.

Energy-making practices and their energies

Most commonly, energy is referred to as one homogeneous entity (Pierce & Paulos 2010). Even though the different energy 'sources' and modes of delivery are commonly distinguished, these represent a very narrow framing of the multitude of possible energies. This remains the case when we limit our discussion to the energies produced in the making and supplying of electricity, as is the focus here. For example, while we can distinguish between different renewable energy technologies, such as solar PVs and wind turbines, these labels do not tell us anything about the *material qualities* of the energies produced by these different sources of power. What, if anything, does windy energy bring to practice that sunny energy does not? For most practices, the answer is nothing. When we flick a switch or turn on an appliance we are not able to (and nor do we necessarily wish to) distinguish between the different properties of energy. Energy is simply energy, no matter where it has come from. And yet history provides us with much knowledge of the different material qualities and meanings that have been attributed to energy in the past.

Consider, for example, the electricity 'cures' which were common until at least 1930, such as electric collars and belts. These devices were intended for rejuvenation and attributed to electricity 'magical' and 'nerve-tingling' qualities. They were designed around the notion that energy was an 'invisible tonic that could restore normal health' (Nye 2010: 75). Meanings of rejuvenation, along with particular ways of handling energy and delivering it to the body, were combined, and energy was integral to and integrated into a distinct restorative health practice. Energy-as-material, while physically no different from the energy you or I might use today, was given mystical qualities that had a profound impact on how it was used in practice and what types of practices it was deemed useful for (Nye 2010). Similarly, it is possible to think about the different qualities energies embody today through the processes and practices involved in making them.

Pierce and Paulos (2010: 117) have considered this idea in their research on energy 'as something interacted with and experienced as a tangible thing'. They describe how practices of making energy can enable different ways of knowing energy, and discuss the possibility of 'transforming people's relationships to energy in more engaging and meaningful ways – for designing energy itself to *matter* to people...to shape and amplify this mattering in more sustainable and desirable ways' (Pierce & Paulos 2012b: 68; emphasis in original). This is a tantalising prospect, and one which Maller and I considered in our research with Australian migrants, where we positioned energies and

Micro-generation 141

waters as some*thing* or *things* that are made, and on which distinctive domestic routines depend (Strengers & Maller 2012).

More specifically, Maller and I discuss how locally situated energy and water systems and sources, such as wood-fired or kerosene-fuelled heaters and ovens, or rivers, wells and water bores, produce their own distinctive resource-making practices (Strengers & Maller 2012). These involve technical materials or equipment (donkeys, axes, clay pots), skills about how to 'handle' different water and energy sources, and meanings about how these resources should be treated (boiled, cooled) or stored (in the home, a sealed container or wood shed). These practices materialise energies and waters in different ways across a host of other domestic practices, where they can bring meanings of scarcity or abundance, along with distinct ways of being handled. For example, the availability of firewood or river water, and the effort required to collect, store and 'make' these resources into usable forms of heat or water, can render these materials precious or plentiful.

Taking the idea of energy-as-material seriously does not mean that energies are different in a *physical* sense: a kilowatt hour (kWh) of wind power is the same as a kWh of solar power. However, it does imply that different micro- and macro-generated energies are materially distinct from one another. For example, consider the different energies produced by micro-wind generators and rooftop solar PV panels. The availability of these two energy sources is one obvious material difference. Solar energy is available at the solar peak of the day, whereas wind energy is available when the wind is blowing at a certain speed. The technologies that materialise these two energies constitute another distinction: wind energy is generated through a turbine and its production is clearly visible through the turning of blades and wind strength. On the other hand, solar PV panels might be visible through a smart meter or IHD, which provides information on when the panels are producing power, or through the weather conditions, where a sunny day correlates with available energy. Each technology requires different forms of maintenance and repair, such as cleaning solar PV panels or performing regular service checks, and each might have other 'back-up' or 'mainstream' energies that are drawn on during times when there is no wind or sun.

We can think of these different energy sources as constituting different energy-making practices that householders either perform themselves, or assign to someone or some*thing* else to perform, such as a maintenance company, electricity utility, or the electricity grid itself, which 'steps in' to provide power when windy and sunny energies are not available. Thinking along these lines, we can start to ask questions such as, what qualities and meanings do the energies produced by these practices bring to everyday activity? What does participating in a micro energy-making practice involve, and what is it like? And how do these practices of making energy potentially intersect with other domestic routines oriented around the availability of these micro-energies – routines like doing the laundry on a sunny day, or trying to use less energy in summer (and more in winter)?

Similarly, we can also start to think of large-scale systems of energy provision as distantly situated and geographically dispersed energy-making practices. These practices involve sequences of activity (extraction, refinement, conversion, production, transmission, distribution) in different and sometimes concealed locations. They are performed by groups of specialised experts who are tasked with delivering energy to homes and buildings in usable forms, such as electricity. Unlike water, which is often positioned as a finite or 'natural' resource even in macro water-making practices, these macro energy-making practices position energy as an infinitely producible or unlimited material (Kurz *et al.* 2005) – that is, some*thing* that can be made *ad infinitum*, is available all the time, and is homogeneous in its make-up and characteristics.

Time and space are important in this constitution of energy-as-material. The energy produced and delivered through macro energy-making practices is commonly made somewhere else by someone else on a timescale that is not apparent or relevant as it enters the home (unless overlaid by other strategies such as dynamic pricing – see Chapter 6). Derived from Big Energy systems (to coin a phrase from Sofoulis' (2005) research on 'Big Water'), this energy is largely invisible, immaterial and inconsequential, unless the systems that deliver it break down[2] or other strategies attempt to make energy meaningful to practice once more. Despite being constituted of multiple 'technologies' the electricity system appears here as a single homogeneous technology that produces one homogeneous and inconsequential energy. In short, energy does not matter very much at all. The following discussion develops this conceptualisation in relation to micro-generation, where I consider the qualities and meanings of the energies that micro energy-making practices potentially produce.

Abundant and scarce micro-energies

Studies that have sought to qualify (and sometimes quantify) the existence of a 'link' between micro-generation systems and consumer behaviour provide important clues for thinking about the qualities imbued in micro-generated energies. For example, Keirstead (2007) found that the installation of solar PV encouraged UK households to reduce their overall electricity consumption by approximately 6 per cent and shift demand to times of peak generation. An Australian survey of solar PV users found that 50 per cent of respondents were taking energy conservation measures after this technology was installed, such as switching off lights, only turning on the heating and cooling when people were in the home, and turning off stand-by appliances (ATA 2007). Similarly, Roy et al.'s (2008) survey of micro-generation 'pioneers' in the UK found that three-quarters of people using micro-generation heat say that they are more aware of their energy use, and make more effort to save energy than they did before. These and other studies suggest that micro-generation somehow engages householders in a range of energy-saving actions (see Chapter 5), or in renegotiating when everyday practices are performed (see Chapter 6) as a result of it being materially present.

Bahaj and James (2007: 2124) support this idea when they highlight the visibility of solar PV systems 'and the direct and clear coupling between the resource (sunlight) and the level of power generation'. They argue that solar PV systems' visibility makes this technology 'one of the best in terms of raising understanding of energy use'. However, Bahaj and James (2007: 2136) are also sceptical about this connection, arguing that expectations that the visible installation of micro-generation will 'keep reminding occupants of the link between energy generation and consumption ... [is] somewhat wishful thinking'. Despite some early success in reducing energy demand, their study found that consumption returned to the previous level within a year as householders adopted a 'proliferation of consumer electronic devices', notably large screen televisions and computers with 'always on' Internet connections, and additional freezers (Bahaj & James 2007: 2133). Thus, as with energy feedback (Chapter 5), the provision of micro-generated energy might both engage householders in a limited number of energy-saving actions defined in relation to 'saving energy' and allow new – and higher – expectations and energy-intensive practices to emerge.

A further possibility is for micro-generated energies to embody meanings of abundance, along with 'rights' and entitlements to this self-produced and renewable resource. There was some evidence of this in Abi-Ghanem and Haggett's (2010: 158) study of solar PVs in UK homes. They describe a group of 'opportunistic users' who justify additional power usage, such as running the clothes dryer when they otherwise would have used a clothes line, because they are making their 'own' energy. Similarly, Juntunen's (2011) study of small-scale renewable energy technologies in Finnish summer cottages found that the increased availability of supply brought the possibility of adding new appliances to householders' cottages, much as the automation strategy of DLC can legitimise the further use of air-conditioning. This might lead us to conclude that, like a green 'traffic light' on an IHD (see Chapter 5) or the assignment of energy management to someone or something else (see Chapter 7), the availability of micro-generation, and the sense of ownership or entitlement over the energies it produces by the people who 'make' it, might justify and legitimise participation in practices-that-use-energy that otherwise would not be performed. Just as micro-generation can position energies as scarce and precious, so too can it position them as abundant, 'free' and 'owned'.

From this brief discussion we can conclude that micro energy-making practices can produce *multiple* energies with different qualities. Sometimes these energies are positioned as scarce, precious and limited; in other examples they are abundant, freely available and imbued with a sense of entitlement or ownership. In order to understand how different micro-energies become imbued with different material qualities and meanings, we need to consider how these energies are made. Here we find some interesting possibilities for how householders participate in energy-making practices, and what they are enroled in performing when they do.

Home grown and handmade energies

The micro-resource management aspirations for householders who produce their own power are often described using terminology that resonates with the rhetoric of the Smart Utopia. Even the terms 'prosumer' and 'co-manager' – which I have myself used in past research (Strengers 2011a) – contain a rationalist undertone, whereby householders are repositioned as the producers and managers of power supply systems, and where 'production' and 'management' are defined in relation to resource management. However, studies of micro-generation reveal other pleasurable and aesthetically important ways in which householders understand their participation in energy-making practices.

For example, participants in Pierce and Paulos' (2010: 121) study of local energy systems liken practices of making energy to gardening, farming or cooking, where the satisfaction obtained is similar to 'growing your own vegetables' or 'tending to your solar garden'. Participants in Dobbyn and Thomas' (2005: 7) study on micro-generation also describe the 'sheer excitement and pleasure' of 'DIY energy generation', again likening the experience to 'growing your own vegetables' rather than to a primarily technological experience. Further synergies can be drawn with micro-cogeneration systems of heat and power, which are sometimes powered by 'own grown biomass', positioning energy as something that is quite literally 'home grown' (Leenheer *et al.* 2011: 5261).

Pierce and Paulos (2012b) stress the importance of the *work* required for householders to participate in energy-making practices, suggesting that energies come to matter to everyday practice through the labour required to make them. They then argue for the design of energy systems to include *personal effort*, 'in terms of immediate time, skill, effort, engagement, and bodily power' (Pierce & Paulos 2012b: 610). In other words, they advocate energy-making practices in which householders are doing some aspect of the making, managing or maintaining. The authors conclude that this may lead 'to improved quality of lived experience – in terms of pleasure, enjoyment, meaning, satisfaction and so on' (Pierce & Paulos 2012b: 610–1). Here, the fact that energy has been 'handmade' renders it a limited and valuable resource – or rather a resource that must be used resourcefully. 'Resourcefulness' becomes a trait that emerges and is reproduced through tangible experience and familiarity in the making or handling of resources, not brought about through changes in individuals' attitudes, values or opinions (although these have been shown to change as a result of direct experiences with resources) (Strengers & Maller 2012: 760).

In these examples, energy is repositioned as something that is made locally in a limited amount, rendering it more visible, material and imbued with meanings of scarcity (or at least *temporal* scarcity). This can be partly explained by the tight coupling between practices of making energy and practices-that-use-energy, which is most pronounced in studies of off-grid micro-generation. For example, one of the participants in Woodruff *et al.*'s (2008: 315) research with 'green' households using off-grid passive solar systems describes their life as 'like living on a ship', where available supplies must be carefully managed and distributed. Similarly, Chappells and Shove's (2004b: 139) study of UK sustainable housing schemes with microgrids found that households 'developed a distinctive approach to demand management' by arranging their routines around the availability of energy. This involved a range of strategies, such as adjusting their clothing to suit the climatic conditions and taking fewer showers in order to 'flatten' peak load (Chappells & Shove 2004b).

These possibilities point us towards the potentially important integration of and intersection between practices of making and using energy. Where these two suites of practices are tightly woven together, micro-generation may present further opportunities for making energies that matter to everyday practice, in ways that shift and shed energy demand.

Temporal orders and reordering practice

The close relationship between micro energy-making practices and practices-that-use-energy can be understood as a 'complex' of practice, or rather a 'stickier and more integrated arrangement includ[ing] co-dependent forms of sequence and synchronization' (Shove *et al.*

2012: 17). In this practice complex, domestic routines and practices of making energy are closely coupled together in a mutually dependent and co-located relationship. This characterisation allows us to consider the idea that the 'integration, sequence and synchronicity between social practices define, constitute and reproduce the rhythmic ordering of daily life' (Pantzar & Shove 2010a: 19). More specifically, it allows us to explore the prospect that the synergistic relationship between practices of making and using energy has the potential to reorder everyday practice.

Before considering these ideas, it is worth emphasising the relationship between time and space in this practice complex. Micro-generated energies are produced at specific times of the day on particular sites. More energy may be available at particular times of the year if energy sources, such as the sun and the wind, are seasonally dependent. The availability of energy will also depend on the physical location of the micro-generation system, its spatial relationship to other things such as roofs and trees, and its ability to 'capture' a given energy source. As a result, time and space both imbue energy with physical qualities linked to when the sun is shining or the wind is blowing. Space can also increase the tangibility of energy and its ability – or need – to be 'handled' on site, or the need to use other materials (such as roofs, trees, timers) in producing usable energies. These time–space dynamics potentially position micro-energies as a tangible, temporal and limited 'thing' or 'things'.

This represents an interesting contrast to the recent inclusion of ICTs in everyday life, many of which have enabled a decoupling and softening of the time and space constraints of many practices, such as working, communicating and engaging in forms of entertainment (Røpke & Christensen 2012). In Chapter 6 I suggested that this softening of time could create opportunities for CPP to facilitate the shifting of routines to different times of day. Furthermore, smart technologies such as ICTs and home automation can allow these routines to be performed in and from other places, by automating specific activities or facilitating opportunities to leave the home (see Chapters 6 and 7). In contrast, micro-generation, while not strictly an ICT but rather part of an ICT-enabled grid, brings specific spatial and temporal constraints to practices of making energy and to the practices that use these energies. Time and space are coupled and tightened, rather than decoupled and loosened.

These time–space dynamics are integral to the constitution of practices of making and using energy as a practice complex. Evidence of this coupling can be found in studies of micro-generation, which report that householders adapt their routines around the availability of power. For example, Roy *et al.* (2008) describe how some householders in their study adapted their behaviour to make best use of the hot water and heat they're generating. Similarly, Dobbyn and Thomas's (2005) study found that householders with micro-generation systems understood the weather conditions that generated power. For some householders with a wind generator, this meant turning off lights and appliances and limiting showering, baths and laundering to a time when the wind was blowing. Pierce and Paulos (2012c: 634) also refer to the temporal relationship between making and using energy in their description of 'windy day laundering and sunny-day bread making'.

A more detailed example is provided by Jalas and Rinkinen (2012), who have studied these dynamics in relation to domestic wood-based heating practices in which energy (in this case heat) is produced by refuelling fireplaces and boilers. In this example the practices of making and using energies, or making and using different types of heat, are so closely bundled together they arguably constitute the same practice. Drawing on Fine's (1990) concepts of rhythm, sequence, periodicity, synchronisation and tempo, the authors show the temporal orders established by different heating technologies, and refer to the 'speeds' of everyday life that they enable. Like dynamic peak pricing and home automation technologies (see Chapters 6 and 7), Jalas and Rinkinen draw attention to the ways in which heat-making practices can generate 'hot spots' and 'cold spots' of activity (Southerton 2003). They highlight the importance of analysing 'temporal order-making' in energy policy that seeks to spread new, low-carbon heating or energy technologies (Jalas & Rinkinen 2012: 1). Their findings resonate with Hallnas and Redstrom's (2001) vision of 'slow technology', and allude to the possibility of designing energy systems, or rather establishing and circulating energy-making practices, that literally slow us – or consumption – down.

Other research finds that micro-generation technologies themselves, rather than the energies they produce or the practices involved in making them, are integrated into the performance of domestic routines. For example, Dobbyn and Thomas (2005: 7) found that householders can attribute 'living credentials' to micro-generation technologies, where they make specific demands on the practices performed within the home. Resonating with the human-like agencies of home automation technologies discussed in Chapter 7, one participant in Dobbyn and Thomas's (2005: 7) study of micro-generation systems gave her heat pump a name, while another described how it was better to 'work with the house rather than work against the house', referring to the need to keep her domestic practices synchronised with the micro-generation system. In similar ways to those described above, this points towards the possibility of conceptualising micro-generation technologies as materials of domestic practices in their own right, potentially bringing new meanings to everyday practices and reordering routines.

It is clear from this discussion that the temporal and spatial relationship between practices of making energy and practices of using energy is integral to how and when practices-that-use-energy are performed. However, it is important not to romanticise such connections or to conclude that 'small is beautiful'.[3] Like other practice complexes, potential 'links' between practices of making and using energy remain 'emergent, generative and creative' (Pantzar & Shove 2010a: 20). This 'connection' is not a universal or stable outcome of producing one's own power; it is a dynamic and mutually dependent relationship, and it is the householders who make and use energies who will, or will not, sustain it through their recurrent performances.

Furthermore, this practice complex of making and using energy may be particularly loose or vulnerable where householders have access to micro- *and* macro-generated energies, such as where they have access to grid-connected micro-generation. In this scenario macro-energies can simply replace micro-energies in times of temporal scarcity, meaning that everyday routines are no longer tightly woven together with the energies required to perform them. This could undermine the potential for micro-generation technologies to enrol householders in domestic routines oriented around the availability of energy, or to constitute energy as a valuable material of practice.

Energies that matter

The analysis above reveals that the ways in which energies are made matter to practice in different ways: it can imbue energies with different qualities and meanings of abundance and scarcity; it can enrol householders as participants in practices of making energy that they liken to pleasurable activities; or it can constitute a practice complex of making and using energy that reorders temporal routines around the availability of energy. What then, can this analysis tell us about how we could design and locate energy systems to reproduce some of these possibilities?

Pierce and Paulos (2010, 2012c) have answered this question by identifying the characteristics of energy systems that lead to energy's mattering. They suggest that visibility and material tangibility are important characteristics of energy systems, as are seasonality (unpredictable and intermittent generation), and contextuality (situated physically close to those who make and use it). In a very small test case, these authors hypothesise that designing systems to emphasise these characteristics may lead to everyday routines that are oriented towards demand management aims. For Pierce and Paulos (2010: 117; emphasis in original), the solution lies in redesigning 'energy as some*thing* more tangible, more differentiated, and less available'. While this might involve specific technical characteristics, these characteristics are defined in relation to the types of energies they produce (tangible, seasonal and contextual).

In Maller and my research with three generations of Australian migrant households (Strengers & Maller 2012), we propose that a similar set of characteristics are important to the ways energy comes to matter in everyday practice. In particular, we find that energy and water systems that are materially present, convey a sense of scarcity or limit to production, and are diverse in the sources they draw on, make active demands on the everyday practices householders participate in. By 'active' we do not mean that energy is constantly and consciously at the forefront of individuals' minds, but rather that it plays a role in the practices that use it, much as the other materials, meanings and skills of practice do. Importantly, we do not assign to energy-as-material some sort of supreme status that overrides or determines practice; rather we argue that this material can bring meanings of 'not wasting', along with specific temporal orders and ways of handling, to practice (Strengers & Maller 2012).

Putting this research together, we can infer that energy systems that are materially present, tangible or require some degree of handling, as well as those that draw on temporally limited, seasonal, intermittent, finite or scarce sources of energy, are better than those that do not. By 'better' I mean more likely to enrol householders in energy-making practices that position energy as a tangible, temporal (or seasonal) and limited (or scarce and valuable) material. Further, these energy-making practices are more likely to intersect with everyday practice in ways that potentially shift and shed energy demand.

Importantly, it does not necessarily follow that energy systems need to be locally built, or that householders need to participate in energy-making practices in order for energies to matter in practice. Rather, as the strategy of CPP revealed (see Chapter 6), there are other ways of positioning energy as a scarce or seasonal material that requires specific ways of handling. Other strategies, such as energy-saving campaigns, energy feedback, feed-in tariffs, targets, rations or load controls can also be reconceptualised along these lines, that is, as ways of repositioning the qualities and meanings of energy-as-material and its role in practice. However, when combined with energies produced by macro energy-making practices, such strategies may find themselves competing with the meanings of abundance and the always-available temporalities of macro-energies. The ability of such strategies to transform everyday practice therefore becomes at least partially dependent on the ways in which energy is quite literally made. This is a critical point for anyone seeking to 'change behaviour' through these or any other strategies intended to reduce or shift energy demand.

Thinking further along these lines, Maller and I (2013) have previously suggested that the energies householders' have used or made in the past can play a role in current practice as a 'practice memory'. We find that Australian migrants' past experiences with and memories of tangible, temporal and limited energies and waters can lie dormant for many years while these householders have access to Australia's intangible, always available and seemingly unlimited energy and water resources. However, these memories can resurface in times of drought, energy shortage or in response to strategies intended to encourage householders to shift or shed their demand, thereby repositioning energy's or water's role in practice. More specifically, meanings of not wasting, along with the skills involved in handling these 'limited' materials, can resurface across a range of practices that use them. This represents a potential opportunity to resurrect past skills in handling energy. Householders who have experienced energy shortages, lived in countries or conditions where there is a limited or intermittent supply of energy, or have had to 'make' their own energy through, for example wood-heating and wood-cooking practices, may have dormant memories of energy as a material that matters to practice (Maller & Strengers 2013). Such forms of embodied knowledge are ignored or discounted from the Smart Utopia, where expert forms of energy knowledge prevail.

Putting energy aside

The perspective adopted in this chapter shifts the focus away from how to get consumers to passively accept or adopt micro-generation technologies, how these technologies (and the data they provide) might elicit more active forms of participation in energy management issues, or how they might constitute an active 'link' between technology and behaviour purely through their increased visibility. Instead, the implication of my analysis is that the 'success' or otherwise of micro-generation technologies hinges on their ability to reposition the role of energy in practice, or rather their ability to *make energies that matter*, not as a commodity, resource unit or impact, but as a material element of practice. More specifically, I find that tangible, temporal and limited energies potentially reorder temporal routines and bring meanings of not wasting to everyday practices, in ways that may serve to reduce and shift energy demand.

There are several limitations with this analysis. First, it does not necessarily follow that making energies at home will make energy matter in these ways. Indeed, in some cases micro-generation can normalise and reinforce new practices-that-use-energy by positioning energy as an abundant and freely available material that those who make it are entitled to use. One possible argument here is that whether or not energy matters to practice is irrelevant if it comes from 100 per cent renewable sources. Indeed, in this scenario the Smart Utopia's aim of decarbonisation is addressed without any changes to how people use energy. However, this position fails to account for the embedded energy associated with making and supplying energy, the inability of any country – so far – to be able (or willing) to supply its entire population with renewable energy, and the increasing world-wide pressures created by growing energy demand. More worryingly, this perspective may reinforce the positioning of energy as an abundant, free and low-impact material, allowing for its continuing integration into practices-that-use-energy in new or more energy-intensive ways.

Second, this analysis potentially overemphasises the role of energy in practice, while underemphasising the other meanings, skills and materials of domestic routines. In seeking to bring energy back into practice, I may have gone too far, and fallen prey to the energy-centric focus of the smart ontology which I have sought to avoid. It is important then to reiterate that energies never directly determine a particular course of action on their own; they are instead positioned in a co-constitutive relationship with other elements of practice. This means that what makes sense for householders to do when new energies intersect with domestic practices is always changing (Schatzki 2002). Practices related to cooling a home, for example, are rapidly transforming: the use of eaves, verandas, cross-breezes and sprinklers in Australian households is declining, as the air-conditioner increases in popularity (Strengers & Maller 2011). Furthermore, this is occurring even as new energy-making practices and their energies come into the home. The role of energies within this fast moving feast is always highly contingent and variable, and while micro-generated energies might embody qualities of scarcity or meanings of 'not wasting', interpretations of 'waste' are always transient and transforming.

However, putting energy to one side is unfamiliar and uncomfortable territory for energy utilities and policymakers, as it involves paying attention to and seeking to intervene in things that seemingly have nothing to do with energy at all. It means attending to things like hygiene expectations, dietary and cooking trends, housing design, fashion and new technologies – such as wine coolers and game consoles – that shape practices such as cooking, eating, bathing, laundering, house cleaning and entertaining. In the following concluding chapter, I argue that paying attention to these issues involves nothing short of re-imagining what the Smart Utopia is, and what it seeks to perform.

9
Reimagining the Smart UTOPIA: A Conclusion

Smart energy technologies are not some fanciful or futuristic idea. They are here – now. However, they rarely constitute the rational and deliberative tools of many energy utilities' dreams. Instead they represent a diverse array of possibilities, being used to justify leaving the clothes dryer on, re-negotiate expectations of thermal comfort, support existing and create new spiritual experiences, or enrol householders in practices of 'growing' their own energy. The absence of these realities from the Smart UTOPIA means that, while it holds exciting potential, it is a fundamentally flawed vision. In seeking to perform a self-reproducing smart ontology in which human action is framed around the idealised energy consumer – Resource Man – the Smart UTOPIA excludes, ignores or seeks to eradicate the vast majority of human experience and energy's role within it. The fact that this internationally pervasive vision has so far failed to receive significant critical interrogation is of serious concern.

This book represents a deliberate move to disrupt the global smart energy agenda and its narrow conceptualisation of social action and change. In interrogating the Smart UTOPIA I have drawn attention to its self-reproducing ontology and considered what falls outside its scope, as well as what other and sometimes *multiple* realities, smart energy technologies are enrolled in performing. I have delved into the spaces excluded from the smart ontology, developing an alternative ontology of everyday practice to understand how smart technologies and the energies they manifest are being integrated into everyday life, where they are involved in performing and transforming everyday routines. This line of enquiry has taken smart technology into places it often goes to but that are rarely discussed – into the aesthetic, entertainment and domestic domains of everyday experience.

Many of the arguments I have made are already well documented in academic debate – indeed, much has been written about technological UTOPIAs, smart technologies, and critiques of the rational consumer. Similarly, social practice theory is now informing a burgeoning field of enquiry into energy consumption and sustainability issues (Gram-Hanssen 2011; Røpke 2009; Shove *et al.* 2012; Strengers & Maller 2012; Warde 2005). However, the 'newness' of smart technology and the extensive hype surrounding its ongoing 'deployment' has meant that many researchers, policymakers and energy utilities have failed to move beyond its UTOPIAn aspirations and make connections to these bodies of work. Rather than focusing on broader cross-cutting similarities in global smart technology developments, they have focused on small differences in 'unique' trials and experiments. It is therefore critical to continue the task this book has begun in reimagining the global fascination with smart energy and its potentialities, as well as the meaning of 'smart' itself.

My use of the word 'imagination', with its fictional connotations, is deliberate. I want to highlight the creative and contingent processes involved in evoking different worlds, and to allude to sensations and experiences that are not based on 'fact' or 'hard data'. In arguing for a reimagination of the Smart UTOPIA I am implying that it is *already* imagined, and that is underpinned by a series of fanciful assumptions. In doing so I call into question the cost-benefit analyses, the demographically and statistically representative research, and the quantitative data that underpin the Smart UTOPIA's fact-like predictions. I call into question the Smart UTOPIA itself.

In this concluding chapter, I want to do four imaginative things. First, to reimagine a world without Resource Man; second, to imagine the new possibilities for smart strategies discussed in this book; third, to consider other disciplinary and interdisciplinary resources that could be drawn upon to reimagine the Smart UTOPIA; and fourth, to envision a Smart UTOPIA that might not involve anything smart or UTOPIAn at all.

Putting Resource Man to bed

Some energy consumers fit the aspirational characterisation of Resource Man. Many, however, do not. Even where Resource Man does exist, he may embody other roles and relations to energy that are not captured by this conceptualisation, and these may serve to undermine the demand management ambitions intended for this consumer. I therefore conclude that it is time to put Resource Man to bed.

A fundamental problem with Resource Man demonstrated throughout this book is that he is inadvertently enrolled in consuming more resources. More specifically, his 'smart lifestyle' involves the establishment of new electricity-enabled ways of cooling, heating and securing bodies and homes, as well as more energy-intensive ways of eating, entertaining, working and playing. This connects with a much longer history of the role that electricity and domestic technologies have played in increasing expectations of comfort, cleanliness and convenience (Forty 1986; Schwartz Cowan 1989; Shove 2003). Many energy utilities, appliance and home automation companies have an active interest in promoting these new or enhanced 'smart' practices, and this has ramifications not only for how and how much electricity is consumed, but also for the embodied energy and materials that go into making smart stuff, and the social and ethical issues that arise when sourcing the materials required to make it. As such, serious questions remain about the sustainability of this global vision, in terms of both the energy required to realise it and the environmental impacts of providing and consuming the energy needed to perform it.

Resource Man's rational, informed and technology-savvy behaviours constitute a second major problem with this conceptualisation, primarily because they are not the routes by which most energy is consumed in the home. However, as the provision of data and technology are the only means by which Resource Man is understood to operate and change, this vision simultaneously excludes all other ways in which practices are already and always changing, as well as a range of other possibilities for intervening in these processes of change.

A third problem with Resource Man is the relationship he necessitates and perpetuates, where by householders are positioned as 'learners', who require education, information and technologies from energy utility 'experts' to assume their resource management role. This linear model overlooks and undermines the practical knowledge householders already possess in managing energy in their everyday lives, and precludes a range of other participatory relationship possibilities between the providers and consumers of power. Even in 'prosumer' and co-managerial relationship models, the relationship between energy providers and consumers is defined and contained in relation to energy production and management.

These and other cracks in Resource Man's character provide the grounds to question whether he should remain the object of the energy industry's UTOPIAn aspirations. But if not Resource Man, then who, or what, should we have in his place? One answer might be not to reject Resource Man altogether, but to diversify his character, to explore his emotional side, to consider the multiple personalities he embodies, and to take his contradictions seriously. However, while this potentially extends Resource Man's possibilities, it maintains a commitment to the Attitudes–Behaviour–Choice (ABC) model (critiqued by Shove 2010a) by focusing our attention on the changing attributes and idiosyncrasies – to the 'other factors' – of Resource Man's character.

Do we then need another suite of UTOPIAn super-heroes, such as Domestic Woman or Comfort Kid, to replace or extend the vision for smart energy consumers? Surely these characterisations run the same risk as Resource Man, where the goal could easily remain one of serving these characters' 'needs' through improved domestic management and comfort control (data and technology). In this way, these characterisations might simply reproduce the smart ontology, rather than critically interrogate and attempt to renegotiating meanings of domesticity and comfort.

For these reasons, it may be time to depart from consumer characterisations altogether. This does not mean that people are irrelevant. Indeed, this book has revealed how people play an integral role in performing and reproducing everyday practices, and in integrating smart technologies and strategies within them. However, extending beyond the category of the consumer allows us to explore other interesting phenomena and generate other potential categories that have relevance for energy demand reduction. For example, we might ask what constitutes 'pleasure' and how this concept is changing, or could change, with the emergence of smart energy technologies. Similarly, we might turn to concepts of culture, ethics, ritual or routine to help understand how smart energy technologies are integrated into everyday life, and how energy is consumed. My approach has been to look to everyday practice as the site where energies and smart energy technologies become implicated in processes of continuity and change. Exploring these possibilities has developed and expanded the potential for smart strategies in everyday life, the opportunities of which I now turn to.

Reimagining smart strategies

Reimagining the smart strategies intended for the Smart UTOPIA involves fundamentally rethinking how these strategies 'work'. As should now be abundantly clear, in the smart ontology they work through unproblematic adoption and rational use by the ideal energy consumer – Resource Man. Smart strategies are the tools of Resource Man's trade, allowing him to measure and make decisions about his energy use, and/or to automate energy management, production and consumption on his behalf. These strategies provide Resource Man with the data he needs to make informed decisions, or designate the 'doing' of energy management to smart technology. In contrast, an ontology of everyday practice proposes that change takes place in and through householders' participation in everyday practices. Change occurs when the elements of practice (meanings, materials, skills) realign in one practice or across a bundle or complex of practices (Shove *et al.* 2012). Smart strategies can disrupt, renegotiate or shift practices-that-use energy, but their ability to do so is always contingent on those who perform these practices, and on the current configuration of elements in circulation.

Shove and Walker's (2010) analysis of London congestion charging provides a practical example of how this plays out in practice. They contend that it would be unhelpful to locate Londoners 'as either victims or beneficiaries of congestion charging' (Shove & Walker 2010: 475). Instead, they argue that Londoners 'responses and reactions constitute the scheme itself', suggesting that it is through the reproduction and transformation of Londoners' practices that the scheme either succeeds or fails (Shove & Walker 2010: 475). Analysing this 'intervention-in-effect', the authors contend that congestion charging is 'an unstable, dynamic and emergent outcome of the way in which constituent elements of London life (cars, bikes, information systems, data, regulation, time, destination and attendant practices) fit together' (Shove & Walker 2010: 475). Like the congestion charge, how smart strategies *work*, and what constitutes these schemes, depends on the practices householders perform, or do not perform, when they encounter them.

Energy feedback, for example, has a minimal impact on everyday practice because it is but one form of feedback among many in everyday life. In Chapter 5 I demonstrated how practices, and particularly domestic practices, are closely mediated by other forms of everyday feedback, such as social feedback from other household members, friends, colleagues and advertisers about whether a particular task, such as doing the laundry, has produced adequately clean or sweet smelling clothing; material feedback from kettles, thermostats and washing machines informing people when and how they should be used; and embodied sensory feedback, such as whether the house 'feels right' or the toilet smells.

Householders might not make rational cost-benefit analyses based on the data provided through energy IHDs and website portals, but this does not mean they are not continually monitoring their practices; nor does it mean that practices cannot or do not change without this information. This observation represents challenges as well as opportunities for advocates of the Smart UTOPIA. How, for example, can we begin to think about reorienting the everyday feedback householders draw upon? Importantly, seeking to intervene in existing forms of everyday feedback does not necessarily involve providing householders with information on a screen. A passively designed house can also provide important forms of 'feedback' about when to open or close windows, blinds or doors in order to maximise thermal comfort.

Another line of enquiry might be to ask how energy feedback can become more meaningful to everyday practice. One answer here is that for energy feedback to matter, energy must first matter to practice in ways that it currently does not. Two ways in which energy can come to matter, albeit for specific periods of time, are through the dynamic pricing strategy of CPP and the provision of micro-generated energy. In Chapter 6 I argued that CPP's current effectiveness in reducing peak demand can be understood in terms of its ability to reposition the meanings of energy in practice, particularly in relation to household cooling, during a defined time period. Similarly, Chapter 8 demonstrated how the ways in which energies are made change their qualities and the subsequent meanings and skills they bring to and reconfigure in practice. Further, I suggested that practices of making and using energy can combine to form a practice complex, in which the temporal and spatial dimensions of micro-generation tie in to everyday practice and routines.

These analyses open up the possibility of *making energies that matter* for how everyday practices are performed. One potential route is to focus, quite literally, on the characteristics, design and location of how and where energy is made. More specifically, energy systems that position energy as a tangible, temporal and limited material are more likely to enrol householders in achieving the aims of the Smart UTOPIA. However, this is not simply an endorsement of Schumacher's (1999) 'small is beautiful' philosophy. Macro-generation systems and other strategies can also reposition the qualities and meanings of energy in practice, as CPP clearly demonstrates.

This analysis opens up a range of other opportunities for disrupting the temporal rhythms and routines of everyday life in ways that potentially reduce peak electricity demand and energy consumption. In addition to pricing signals, which can reposition the meanings of energy in practice, other strategies of disrupting or reconfiguring routines become possible – such as alerts, rations, restrictions and limits during peak periods. Water restrictions, water targets, congestion charges and bushfire alert systems are related possibilities that seek to shift 'normal' practices into a temporarily negotiable space.

Similarly, we can also think of home automation technologies as important material disruptors of routines, as well as potential legitimisers of practice. In Chapter 7 I challenged the Smart UTOPIA's intention for home automation technologies to passively automate practice, finding instead that the assignment of energy management to technology can justify, and even increase, expectations associated with practices-that-use-energy. However, these are not the only realities that automation technologies can perform. They can also renegotiate and interfere with everyday routines, enacting scenarios that are different from those currently intended for these devices in the Smart UTOPIA. There are good reasons to be cautious with home automation technologies, as well as some reasons to be positive about

162 *Smart Energy Technologies in Everyday Life*

its potential. One promising possibility is to imagine (and potentially design) ways that these technologies might 'act back' or 'act up' in practice, reorienting what it means to be comfortable, be a good host or achieve pleasance in ways that reduce or shift energy demand.

There is great scope for exploring and extending these possibilities in future work. In the following section, I consider what role academic researchers can play in this task of imagining and performing alternate realities of and for the Smart UTOPIA.

Imagining alternate realities

Researchers play a critical role in reproducing the smart ontology, as well as disrupting and imagining alternatives. In this section I wish to draw attention to some of these disruptive routes, and to the potential pitfalls involved in bringing additional theoretical resources to the task of reimagination. Importantly, I do not wish to generate an exhaustive list or call for an interdisciplinary bucket in which we can throw as many perspectives as possible. All this would achieve is an incomprehensible and incompatible mismatch of conceptual paradigms. Rather, following the tradition of some science and technology scholars and social theorists, I wish to suggest that *multiple* ontologies are necessary to imagine *different* realities for the Smart UTOPIA (Blaikie 1991; Law 2009; Shove 2011).

Shove (2010a: 1279) has previously made this point by arguing that different paradigms of social change are like 'chalk and cheese', with an unbridgeable gulf between them. In regard to the Smart UTOPIA, this observation suggests that researchers, policy-makers and energy analysts should represent different paradigms as just that – different – rather than attempt to combine them. This implies not a rejection of interdisciplinary research, but a resistance to the convergence of different disciplinary perspectives in order to create one unanimously agreed problem and selection of solutions (Evans & Marvin 2006; Strengers 2012b). It involves recognising that different disciplines define problems in different ways, leading to sometimes dramatically different 'solutions' (Shove 2010a), including different understandings of how solutions work and who or what makes them work.

For example, the smart ontology presents peak demand as a problem that is best solved through information and technology targeted at individual consumers. On the other hand, the ontology of everyday practice frames the problem of peak demand as a symptom of transforming household practices, particularly those involving indoor cooling, although this might change in the future. Attempting to intervene in the peaky trajectory of cooling practices involves paying attention to building design, standards and orientation; the changing meanings (and marketing) of indoor health and comfort; international thermal comfort standards; and a corresponding decline in other skills and materials used to stay cool (Strengers 2012a).

Negotiating these distinctions is not always easy, not least because the smart ontology is pervasive and ubiquitous, making it sometimes difficult to know when we, as researchers, are still performing it. This is especially problematic because research is often funded by bodies that are firmly grounded in the smart ontology. Nonetheless, there are several important clues researchers can look for to assess whether their own practices are located within or outside the smart ontology. First, we can ask ourselves whether we are inside or outside the places and spaces typically reserved for social scientists. Is our research located at the end of the supply chain, without reference or theory as to how that supply chain is interacting with everyday life? Second, we can check who or what is the focus of our enquiry. Are we focused on the ABC model, or on understanding individuals' attitudes, behaviours and choices? And third, we can ask ourselves what theories of change we are drawing on. Are we viewing change as a linear or causal process? Are we treating humans and technologies as distinct and separate processes of change, or as part of a one-way relationship whereby technologies simply influence and act on the people who use them? If we are answering yes to one or more of these questions we may find that our own practices are still performing the smart ontology.

How then might researchers begin to perform other realities? As I suggested earlier, we might focus our attention not on people or consumers, but on places, practices, publics, politics and infrastructures. These diverse lines of enquiry have led scholars to study the ethnography of infrastructure (Star 1999), the changing dynamics of comfort, cleanliness and convenience conventions and their energy implications (Shove 2003), the sensory and experiential flows of energy in the home (Pink 2012b), the connections between infrastructure and consumption (Van Vliet *et al.* 2005), the design of energy systems (Pierce & Paulos 2010), and the history of electricity provision and its associated expectations (Hughes 1983; Nye 2010; Trentmann 2009). Other researchers enrolled in reimagining the smart UTOPIAn agenda are considering the modes of material participation different smart devices engender (Marres 2012a), and the types of relationship arrangements new energy infrastructures can constitute and arrange (Chappells & Shove 2004b; Marvin *et al.* 1999; Strengers 2011a).

Homing in on everyday practice, we find similar promising opportunities for future research, such as understanding how practice elements from the past can be resurrected and reintegrated into current practices (Maller & Strengers 2013; Shove & Pantzar 2005b), how energy and its infrastructures are materialised as part of practice (Strengers & Maller 2012), how the practices of policymakers and electricity utilities intersect with domestic routines (Shove *et al.* 2012), or how new or modified practices involving automation and ICTs are emerging and reordering everyday routines (Røpke & Christensen 2012; Røpke *et al.* 2010).

Additionally, there are many other smart strategies to explore, such as the increasing integration of electric vehicles and the gamification of energy consumption. Looking further afield there are opportunities to extend this research agenda into other sectors where the smart ontology is beginning to proliferate, such as the water (Hauber-Davidson & Idris 2006), transport (Climate Group 2008), and health sectors (Lindsay 2010; Purpura *et al.* 2011). Similarly, we might explore other bundles or complexes of practices, such as those of energy providers, policymakers, housing developers, system engineers, appliance manufacturers, computer programmers and advertisers – practices which necessarily intersect with those performed in the home, or arguably have a vested interest in performing the smart ontology and its commitment to Resource Man.

Making this research meaningful to policymakers and energy providers will not be easy, particularly when they are, understandably, deeply committed to the smart ontology and its methods of reducing consumers to targetable, controllable and manageable segments. One way of getting around this problem is by subscribing to a methodology of counting and segmenting practices and their changing dynamics, rather than people and their changing attitudes. A group of UK researchers have recently done just that in a major project on residential water consumption, where they sought to map the patterns of water-using practices in households across the South and South East of England (Browne *et al.* 2013; Pullinger *et al.* 2013). As the authors themselves note, this process is not without its challenges, but it nonetheless represents a promising method of enquiry that attempts to bring new ontological resources to dominant practices of 'knowing', which require some*one* or some*thing* to be counted.

Finally, it is important to note that if we subscribe to an ontology of everyday practice, it is not possible (or necessarily desirable) to provide a prescriptive set of disciplines, methods, strategies or practices that will achieve the aims of the Smart UTOPIA. This is because practices have 'emergent and uncontrollable trajectories' and the potential to intervene in them is always 'complicated and qualified' (Shove & Walker 2010: 475). This does not mean that we should do everything, or nothing. But attempting to do *something* means stepping outside the smart ontology and reimagining the Smart UTOPIA; a task that in itself is intended to disrupt the practices that this vision is currently enrolled in performing, opening up alternative possible realities and futures.

Reimagining smart and UTOPIA

A final contentious implication of this reimagination agenda involves recognising the limits of the Smart UTOPIA, and imagining realities that are neither smart nor UTOPIAn. This might involve turning our attention to the myriad practices that are not 'smart' – in the sense that they do not involve ICTs or quantifiable data. Interviews with householders reveal the mundane dynamics and negotiations of everyday life, where using the hairdryer is about getting those annoying bits of hair to sit flat, and washing in hot water is about ensuring clothes and clean, fresh and hygienic. The meanings of smart are rarely featured in this dialogue, and where they are they are often discussed in relation to a defined suite of actions and activities associated with saving energy (Strengers 2011c). Yet these snippets of everyday life, while seemingly small and insignificant, reveal important transformative potentialities that are excluded from current aspirations for a smart world.

As I have already suggested, everyday insights might focus our attention on the changing meanings of health and cleanliness

(Shove 2003; Strengers & Maller 2011), the changing practices of house design and construction, or the paradigms and practices of thermal comfort (Brager & de Dear 2003; Chappells & Shove 2004a). These are not areas that energy providers or governments have traditionally ventured into, partly because they are solely focused on the provision of quantitative data or ICT to address energy-related issues and enact change, and partly because these are the terms and conditions through which they have the 'right' to 'influence customers'. And yet they are areas of enquiry that are just as important, if not more so, than those that fall within 'smart' containment lines.

There are of course risks and problems with imagining alternative UTOPIAs, particularly if they reject everything deemed smart. For example, we might generate an idealistic vision that is equally problematic, such as a world of self-sufficiency that is free of ICT and 'excessive' consumption. Clearly, smart technology is becoming a ubiquitous part of our lives, even in countries with limited and unreliable electricity supply. Imagining a world without it seems as unrealistic as imagining a world in which we all use it in a rational and deliberative way. These are traps and gaps of which researchers, energy providers and policymakers reimagining the smart agenda need to be aware.

Still, it is possible – and necessary – to reimagine and perform different meanings and definitions of 'smart' and idealised visions of 'UTOPIA'. One approach might be to imagine futures where smart living embodies opportunities for slow time, family time or down time in the flow of everyday activity (Hallnas & Redstrom 2001; Southerton 2003), such as where variable pricing, micro-generation and home automation enable flexible and negotiable routines. Another might be to imagine houses that enable adaptive cooling and heating practices, where 'smart' does not mean the provision of automated and rationally managed, centrally controlled cooling and heating systems, but rather passively designed housing infrastructures where windows, doors, blinds and plants become important 'materials' of heating and cooling practice (Strengers & Maller 2011).

Resisting the growing tide of smart UTOPIAn 'evidence' will not be easy, especially as the stakes – Resource Man and his associated tools – are so high. Many utilities, governments, researchers and other businesses are now thoroughly invested in seeking to perform Resource Man and his associated reality of rational and measured energy management. And yet this task has never been so essential as the pervasive smart vision continues to permeate not only the energy sector, but all realms of social life, perpetuating a fundamentally unsustainable vision of the future. Reimagining a Smart UTOPIA grounded in the mundane realities of everyday life, as pursued in this book, is one alternative that disrupts this dominant agenda. Here, doing the dishes, washing the laundry, cooling the home and cooking a meal form the basis of social order and change, and are the sites where the Smart UTOPIA's aims will or will not be achieved. If I were to reimagine the Smart UTOPIA, this is where I would begin.

Glossary

Critical peak pricing (CPP): A dynamic pricing program where up to 12–20 critical peak 'events' are called throughout the year, during which time the price or electricity rises substantially (20–40 times the usual rate). CPP periods usually last for 2–5 hours and often coincide with hot summer or cold winter days when there is high peak electricity demand. CPP is offset by lower rates during off-peak periods. Events are communicated to households via a range of ICTs, including email, SMS and/or IHDs. Also known as dynamic peak pricing.

Critical peak rebate (CPR): Same as CPP only customers receive a financial rebate when they reduce their electricity consumption during critical peak events instead of being charged a higher tariff during these times. Also known as peak time rebate or dynamic peak rebate.

Demand management: The modification of consumer energy demand through various methods, such as dynamic pricing, education, automation and other incentives or disincentives. Also known as demand-side management.

Demand response: Changes in end users' electricity demand relative to their normal consumption patterns in response to load control technologies, changes in electricity prices or other smart technologies and strategies.

Direct load control (DLC): The remote control of energy appliances, especially high energy-consuming appliances, during periods of peak demand. The control of these devices can be enabled with and without a smart meter, through radio frequencies and powerline communication systems.

Dynamic pricing: Electricity pricing tariffs that change during particular times of the day or in response to changes in energy demand. Generally refers to time-of-use tariffs, critical peak pricing (CPP) and real-time pricing. Also known as time-based pricing and variable pricing.

Glossary 169

Electric vehicle (EV): A vehicle that uses electric or traction motors for propulsion. EVs are powered by stored energy originating from an external power source, an on-board electrical generator, or directly from an external power station. EVs, particularly electric cars with batteries that are charged by the electricity grid, are considered an important technology of the smart grid that may enable opportunities for demand management.

Everyday practices: Practices that are routinely performed in the course of everyday living. Refers to practices performed in or around the home, such as cooking, showering and laundering.

Home automation: The increased automation of household appliances, devices and building features, particularly through electronic means. Generally refers to automatic or semi-automatic lighting; heating, ventilation and air-conditioning services; entertainment systems; climate-controlled windows and doors; and security or surveillance systems.

Home energy management (HEM) system: A device, platform or application that provides energy consumption data and information. Common HEM systems include in-home displays (IHDs), website portals and mobile applications.

In-home display (IHD): A device that displays electricity usage information. Most include real-time and historical data for kilowatt hours, greenhouse gas emissions and the costs associated with a household's energy consumption. Some also include gas and water consumption. Also known as a home energy management (HEM) system.

Kilowatt hour (kWh): A unit of energy which is the product of power in kilowatts and time in hours. For example, a heater rated at 1000 watts (1 kilowatt) operating for one hour uses one kWh of energy. Kilowatt hours are units that are commonly used by electricity providers to bill electricity consumers.

Load factor: Defined as the average load divided by the peak load in a specified time period. A high load factor means that energy demand is relatively constant, and consequently energy costs are more evenly distributed over more kilowatt hours of energy output. A low load factor characterises a network with high peak demand, which means a large proportion of the network sits unused for most of the time.

Load shifting: A demand management strategy employed by electricity utilities to smooth out peaks and troughs in demand by encouraging consumers to shift their consumption from peak to off-peak periods through strategies such as dynamic pricing and home automation.

Micro-generation: A small-scale system of electricity provision and/or heat generation used by households and other buildings, such as schools or offices, to generate power on site. Usually refers to distributed or renewable generation, such as solar photovoltaic (PV) panels and wind turbines.

Micro-grid: A localised grouping of electrical loads and sources that is able to operate independently of the centralised grid or in connection with it, depending on the market conditions for electricity.

Ontology: The philosophical study of the nature of reality and being. Used in this book to depict two alternative manifestations of ways of being in and understanding the world and its realities: the smart ontology and the ontology of everyday practice.

Ontology of everyday practice: An understanding of social reality where all human action and social change is mediated by and through participation in routinely performed practices. Discussed in relation to the everyday practices performed in and around the home.

Peak electricity demand: The period of time where demand for electricity is at its highest. The term generally refers to network or critical peaks, which occur on very hot or cold days when the network is working at capacity or when supply is unable to meet demand. Peak demand can also refer to daily peaks that occur during the morning and/or early evening.

Photovoltaic (PV): A form of electricity generation that employs solar panels composed of a number of solar cells containing photovoltaic material that convert solar radiation into electricity.

Plug and play: Commonly used in consumer electronics industry to refer to the scenario where the only involvement or knowledge consumers require to operate a technology is to 'plug it in'.

Following this simple step consumers can begin 'playing' with the technology.

Practice: A routinised form of socially shared action comprising three interconnected elements: materials, meanings and skills, which constitute a practice entity. Practices are carried out by people, who sustain, modify or transform them through regular performance. Also referred to as social practice and everyday practice.

Prices-to-devices: A feature of the smart grid that allows electricity prices to be sent to decision-making energy appliances that turn on or off in response to these signals.

Programmable thermostat: A heating or cooling thermostat that can be programmed to come on or turn off at certain times of the day. Distinguished from a smart thermostat by its inability to communicate to other devices or third parties outside the home.

Prosumers: A consumer of energy who is also a producer of energy. The term was originally coined by Alvin Toffler, who predicted a blurring of the relationship between producers and consumers.

Real-time pricing (RTP): A pricing tariff that passes on the hourly market rate of electricity to consumers. Typically used for larger commercial customers rather than small businesses or residential customers.

Resource Man: The ideal smart energy consumer. A technologically interested, gendered and highly informed micro-resource manager who is involved in managing his own consumption as well as assigning control of this management to energy utilities and smart technologies.

Set-and-forget: A feature of the smart grid that allows householders or electricity providers to automate smart appliances to come on or off at specific times in response to pricing signals or to achieve demand management objectives.

Smart appliance: An appliance characterised by two-way communication, and which can be remotely controlled or pre-set. Smart appliances can respond to dynamic pricing tariffs and other demand management incentives.

Smart grid: An electricity grid that uses information and communication technologies (ICTs) to gather and act on information in order to efficiently, reliably and sustainably manage the demand and supply of electricity. Generally encompasses a broad suite of technologies and strategies including smart meters, automation technologies, dynamic pricing, electric vehicles and micro-generation. There is currently no internationally accepted definition of a smart grid and most are still in pilot stages.

Smart metering: A form of electronic metering which replaces the traditional manually read electricity, water or gas meter as a way of measuring the quantity supplied to or produced by a residence or business. Definitions of smart metering vary, partly because this device's functionality is still being defined. Generally speaking, a smart meter is characterised by at least half-hour data logging capability and two-way communication functionality (between the metered property and the utility).

Smart ontology: The understanding of social reality underpinning the Smart Utopia, whereby all human action and social change are mediated by information communication technologies (ICTs) and data.

Smart thermostat: A heating or cooling thermostat with two-way communication functionality that can be remotely programmed to turn on or off at specific times of the day or in response to price or other demand management signals. Automated responses to these signals can usually be overridden by the users or owners of this device. Also known as a two-way thermostat.

Smart Utopia: The international vision for smart energy technologies, in which the social and environmental problems facing the electricity sector are solved by data and technology. The Smart Utopia builds on technological utopian ideas from the late nineteenth and early twentieth centuries.

Time of use (TOU) tariffs: A form of dynamic pricing that involves different tariffs (two or more) for different times of the day. These times represent the average peaks over a 24-hour period but do not represent critical or network peaks in demand.

Notes

1 Introducing the Smart Utopia

1. Peaks in demand, which may only occur for one week of the year, result in the price of electricity spiking in the energy market as less efficient and older generation plants come online to meet demand. They also require significant investments in electricity generation, distribution and transmission infrastructure which is unused for the majority of the year (Faruqui & Palmer 2011). Peak demand is growing in many western nations with increases in residential air-conditioning and the changing technologies and practices associated with consumer electronics. In Australia, for example, peak demand is growing at a faster rate than average demand (AEMC 2012), resulting in new infrastructure being built to cope with those few days of the year when it is needed. The economic implications of these investments are significant for both utilities, in the form of capital costs, and consumers, to whom these costs are eventually passed.
2. Direct load control (DLC) refers to the remote control of large appliances by energy utilities during periods of peak demand or grid stress. It usually applies to air-conditioning and hot water systems in households, which are cycled on and off for short periods of time.
3. Micro-generation is a term used to describe small-scale systems of electricity provision (generally up to 50 kilowatts) and/or heat (up to 45 kilowatts thermal), which are used by households or community buildings (such as schools or office buildings) to generate power on site.
4. A micro-grid is 'a collection of geographically proximate, electrically connected loads and generators' (Platt *et al.* 2012: 186).
5. Gamification is the use of game thinking (competition, play, fun, achievement, fulfillment) in applications designed to improve customer engagement. In the energy sector, it is being used to encourage householders to compete with each other to reduce energy. One example is Power House, an energy game that connects home smart meters to an online multiple player game with the goal to improve home energy behaviour (Reeves *et al.* 2012).
6. Load factor is defined as average electricity demand divided by peak electricity demand over a period of time. A high load factor characterises infrastructures where electricity demand is relatively constant. A low load factor refers to infrastructures with a large disparity between average and peak demand, meaning that more of the infrastructure is used less of the time (*e.g.* it is only used for peak periods).
7. Smart thermostats are used with heaters and/or air-conditioners to communicate with and be controlled by energy providers or demand

response systems. They are also known as two-way thermostats and programmable thermostats.
8. The Association of Home Appliance Manufacturers (AHAM) defines a smart appliance as 'a modernisation of the electricity usage system of a home appliance so that it monitors, protects, and automatically adjusts its operation to the needs of its owner' (in Hamilton *et al.* 2012: 409–10).

2 Imagining the Smart Utopia

1. Any references to this book are from More, T. 2005 *Utopia*, Barnes & Noble, New York.
2. There are a growing number of websites and community groups devoted to the task of 'stopping' smart meters being installed on properties. See, for example, http://stopsmartmeters.org/.
3. The 1930s Homes of Tomorrow represented ideals of the future rather than contemporary reality. Appliance manufacturers General Electric and Westinghouse built their own display houses of tomorrow, representing a consumerist orientation towards increasing appliance and electricity use (Horrigan 1986).
4. 'Set-and-forget' refers to smart appliances that householders can set on a timer to turn on and off at certain times of the day. This enables householders to 'forget' about these appliances once the initial setting has been done. Similarly, 'prices-to-devices' refers to appliances that can be set to turn on or off in response to higher or lower dynamic electricity tariffs that charge more or less for electricity during different times of the day.
5. Mini-Me is a character in the second and third Austin Powers movies: *Austin Powers: The Spy Who Shagged Me* and *Austin Powers in Goldmember*. He was a clone Dr Evil made of himself. Mini-Me was identical to Dr Evil in every way, but he was one-eighth his size. Relating this concept to the water sector, Sofoulis (2011: 805) writes that 'Mini-Me-ism is prominent in rationalist approaches, where water authorities conventionally assume that users will (or ought to) think just like they do, and value the kinds of rational and technical knowledge that water experts consider important. This assumption leads to research, educational and consultative strategies whose effectiveness may be limited to that population minority who respond well to quantitative technical and economic resource consumption data, leaving those with other orientations and motivations unengaged.'

3 Resource Man

1. In 2009, the OECD average percentage of bachelor and higher level degrees awarded to women in these or related professions was engineering, manufacturing and construction (26.3 per cent); computer science (19.2 per cent) (NCES 2012: 625). I was unable to source comparable figures for the social science discipline of economics.

2. See https://www.bchydro.com/youraccount/teampowersmart/Join.do.
3. See http://www.p3international.com/products/special/P4400/P4400-CE.html.
4. Described as 'a consumption metering piggy bank designed to sensitize kids to energy costs associated with running electronics devices': http://www.core77.com/greenergadgets/ientry.php?projectid=50.
5. See http://twitter.com/#!/tweetawatt.
6. See http://opower.com/.
7. See http://www.ecomagination.com/.
8. Akrich (1992) is referring here to a cumulative collection of meters, such as those included in a large-scale smart metering project.

4 Energy in Everyday Practice

1. Shove (2010a; 2010b) has made a similar argument in relation to climate change policymaking and its over-reliance on the Attitudes, Behaviour, Choice (ABC) model.
2. Home weatherisation involves weather-stripping doors and windows and sealing joints and cracks to keep out draughts. It is considered a relatively inexpensive energy conservation strategy with quick paybacks (Wilk & Wilhite 1985).

5 Energy Feedback

1. See http://www.alertme.com.
2. See http://www.currentcost.com.
3. See http://www.ecometer.com.au/.
4. See http://www.theowl.com.
5. See http://www.clipsal.com.au/consumer/products/energy_Efficient_solutions/energy_consumption_monitor.
6. See http://www.diykyoto.com/uk.
7. See http://www.tendrilinc.com/platform/connect/.
8. See http://www.energyhub.com/.
9. www.garmin.com.
10. For example, see www.strava.com.
11. See http://www.mapmyrun.com/.
12. See http://www.p3international.com/products/special/p4400/p4400-ce.html.

6 Dynamic Pricing

1. Peaking power plants are generally only run when there is high demand, or peak demand, for electricity. Because they are only run during peak times the cost per kWh to run them is much higher than that supplied by a base load power station.

2. Information-only refers to a type of control group sometimes run in trials of CPP, whereby households receive notification of a CPP event via a range of preferred ICTs (phone, mobile, email, IHD), but remain on a flat electricity tariff. Some companies are now trialling 'information-only' products (also known as a 'peak alert'), which notify householders of an upcoming peak event but do increase or reduce the price of electricity during this time.
3. Approximately half the world's electricity consumers must continuously improvise power and light (Nye 2010: 225)

8 Micro-generation

1. 'Plug and play' is a term commonly used in consumer electronics. It refers to the scenario where the only involvement or knowledge consumers require to operate a technology is to 'plug it in'. Following this simple step consumers can begin 'playing' with the technology.
2. See Nye's (2010) history of American blackouts for a comprehensive account of how energy can rematerialise when the systems that provide it break down.
3. This phrase was made popular by Schumacher (1999) in his 1970s book *Small is Beautiful: Economics as if People Mattered*. It is often used to champion small-scale technologies, such as micro-generation, that are expected to empower people to live less resource-intensive lives.

Bibliography

Abi-Ghanem, D. and Haggett, C. (2010) 'Shaping people's engagement with microgeneration technology: The case of solar photovoltaics in UK homes', in P. Devine-Wright (ed.), *Renewable Energy and the Public: From nimby to Participation*, Earthscan, London, 145–65.
ABS (2011) 4602.0.55.001 – Environmental Issues: Energy Use and Conservation, Australian Bureau of Statistics, Canberra.
Accenture (2010) Engaging the New Energy Consumer: Accenture Perspective-Operational Imperatives for Energy Efficiency, Accenture, Dublin.
Accenture (2011) Revealing the Values of the New Energy Consumer: Accenture End-Consumer Observatory on Electricity Management 2011, Accenture, Dublin.
Accenture (2012a) Actionable Insights for the New Energy Consumer, Accenture, Dublin.
Accenture (2012b) The New Energy Consumer: Balancing Strategic and Operational Imperatives. Reference Guide 2.0, Accenture, Dublin.
Ackermann, M. (2002) *Cool Comfort: America's Romance with Air-Conditioning*, Smithsonian Institution Press, Washington.
AEMC (2011) Issues Paper: Power of choice – Giving Consumers Options in the Way They Use Electricity, Australian Energy Market Commission (AEMC), Sydney.
AEMC (2012) Overview Summary: Power of Choice Review – Giving Consumers Options in the Way They Use Electricity, Australian Energy Market Commission (AEMC), Sydney.
Agbemabiese, L., Berko, K., Jr. and du Pont, P. (1996) 'Air conditioning in the tropics: Cool comfort or cultural conditioning?', in W. Kempton and L. Lutzenhiser (eds), 1996 ACEEE Summer Study on Energy Efficiency in Buildings: Human Dimensions of Energy Consumption, American Council for an Energy-Efficient Economy (ACEEE), Washington.
Akrich, M. (1992) 'The de-scription of technical objects', in W.E. Bijker and J. Law (eds), *Shaping Technology/Building Society*, The MIT Press, Cambridge, Massachusetts, 204–24.
Allcott, H. (2009) *Social Norms and Energy Conservation*, Centre for Energy and Environmental Policy Research (CEEPR), Massachusetts.
Anderson, W. and White, V. (2009) *Exploring Consumer Preferences for Home Energy Display Functionality*, Centre for Sustainable Energy for the Energy Savings Trust, Bristol.
ASHRAE (2004) *Standard 55: Thermal Environment Conditions for Human Occupancy*, American Society of Heating Refrigeration and Air-Conditioning Engineers, Atlanta.

ATA (2007) *The Solar Experience: PV System Owners' Survey*, Alternative Technology Association, Melbourne.

Ausgrid (2012) *Smart Home Update: Energy Analysis*, Ausgrid and Sydney Water, Sydney.

Bahaj, A.S. and James, P.A.B. (2007) 'Urban energy generation: The added value of photovoltaics in social housing', *Renewable and Sustainable Energy Reviews*, vol. 11, no. 9: 2121–36.

Barad, K. (2003) 'Posthumanist performativity: Toward an understanding of how matter comes to matter', *Signs*, vol. 28, no. 3: 801–31.

Beck, U. and Beck-Gernsheim, E. (2002) *Individualization: Institutionalized Indvidualism and its Social and Political Consequences*, Theory, Culture and Society, SAGE Publications Ltd, London.

Bell, G. and Kaye, J. (2002), 'Designing technology for domestic spaces: A kitchen manifesto', *Gastronomica*, vol. 2, no. 2: 46–62.

Bell, G., Blythe, M. and Sengers, P. (2005) 'Making by making strange: Defamiliarization and the design of domestic technologies', *ACM Transactions on Computer-Human Interaction*, vol. 12, no. 2: 149–73.

Berg, A.J. (1994) 'A gendered socio-technical construction: The smart house', in C. Cockburn and R. Furst Dilic (eds), *Bringing Technology Home: Gender and Technology in Changing Europe*, Open University Press, Buckingham, 165–80.

Bergman, N. and Eyre, N. (2011) 'What role for microgeneration in a shift to a low carbon domestic energy sector in the UK?', *Energy Efficiency*, vol. 4, no. 3: 335–53.

Berker, T. (2013) '"In the morning I just need a long, hot shower": A sociological exploration of energy sensibilities in Norwegian bathrooms', *Sustainability: Science, Practice & Policy*, vol. 9, no. 1: 57–63.

Berry, M., Gibson, M., Nelson, A. and Richardson, I. (2007) 'How smart is "smart" — smart homes and sustainability', in A. Nelson (ed.), *Steering Sustainability: Policy, Practice and Performance in An Urbanising World*, Ashgate, Burlington, 239–52.

Berst, J. (2012) 'Smart grid momentum: Think less about people more about devices', SmartGridNews.com, 5 March 2012, http://www.smartgridnews.com

Blaikie, N.W.H. (1991) 'A critique of the use of triangulation in social research', *Quality & Quantity*, vol. 25, no. 2: 115–36.

Borenstein, S., Jaske, M. and Rosenfeld, A. (2002) *Dynamic Pricing, Advanced Metering, and Demand Response in Electricity Markets*, Center for the Study of Energy Markets, University of California Energy Institute, California.

Bourdieu, P. (1977) *Outline of a Theory of Practice*, Cambridge University Press, Cambridge.

Bourdieu, P. (2005) 'Habitus', in J. Hillier and E. Rooksby (eds), *Habitus: A Sense of Place*, 2nd edn, Ashgate Publishing Ltd, Harts, 43–9.

Brager, G.S. and de Dear, R.J. (2003) 'Historical and cultural influences on comfort expectations', in R. Cole and R. Lorch (eds), *Buildings, Culture and Environment: Informing Local and Global Practices*, Blackwell Publishing, Oxford, 177–201.

Brager, G.S., Paliaga, G. and De Dear, R.J. (2004) 'Operable windows, personal control, and occupant comfort', *ASHRAE Transactions*, vol. 4695, no. RP-1161: 17–35.

Browne, A., Medd, W. and Anderson, B. (2013) 'Developing novel approaches to tracking domestic water demand under uncertainty—A reflection on the "up scaling" of social science approaches in the United Kingdom', *Water Resources Management*, vol. 27, no. 4: 1013–35.

Brynjarsdottir, H., Hakansson, M., Pierce, J., Baumer, E., DiSalvo, C. and Sengers, P. (2012) 'Sustainably unpersuaded: How persuasion narrows our vision of sustainability', paper presented to Proceedings of the 2012 ACM annual conference on Human Factors in Computing Systems, Austin, Texas.

Butler, J. (2010) 'Performatice agency', *Journal of Cultural Economy*, vol. 3, no. 2: 147–61.

Caird, S. and Roy, R. (2010) 'Yes in my back yard: UK householders pioneering microgeneration technologies', in P. Devine-Wright (ed.), *Renewable Energy and the Public: From nimby to Participation*, Earthscan, London, 203–17.

Caplan, B. (2001) 'What makes people think like economists? evidence on economic cognition from the "survey of Americans and economists on the economy"', *Journal of Law and Economics*, vol. 44, no. 2: 395–426.

Carey, J.W. and Quirk, J.J. (2009) 'The history of the future', in J.W. Carey (ed.), *Communication as Culture: Essays on Media and Society*, Routledge, New York, 133–54.

Carlsson-Kanyama, A. and Lindén, A-L. (2007) 'Energy efficiency in residences—Challenges for women and men in the North', *Energy Policy*, vol. 35, no. 4: 2163–72.

CEA (2011) Unlocking the Potential of the Smart Grid – A Regulatory Framework for the Consumer Domain of Smart Grid, Consumer Electronics Association (CEA), Arlington.

Challis, C. (2004) A Literature Review of Secondary and Smart Metering Knowledge in Managed Housing, The Energy Savings Trust, London.

Chappells, H. and Shove, E. (2004a) 'Comfort: A review of philosophies and paradigms', Lancaster University, Lancaster, http://www.usp.br/fau/cursos/graduacao/arq_urbanismo/disciplinas/aut0264/Material_de_Apoio/Chappells_Shove_2004_Comfort_Philosophies_Paradigms.pdf.

Chappells, H. and Shove, E. (2004b) 'Infrastructures, crises and the orchestration of demand', in D. Southerton, B. Van Vliet and H. Chappells (eds), *Sustainable Consumption: The Implications of Changing Infrastructures of Provision*, Edward Elgar, Cheltenham, 130–43.

Chappells, H. and Shove, E. (2005) 'Debating the future of comfort: Environmental sustainability, energy consumption and the indoor environment', *Building Research and Information*, vol. 33, no. 1: 32–40.

Climate Group (2008) *Smart 2020: Enabling the Low Carbon Economy in the Information Age*, Global eSustainability Initiative, Brussels.

Control4 (2013) *Residential*, viewed 15 February 2013, http://www.control4.com/residential/.

Cooper, G. (1998) Air-conditioning America: Engineers and the Controlled Environment, 1900–1960, The John Hopkins University Press, Baltimore.

Couldry, N. (2012) Media, Society, World: Social Theory and Digital Media Practice, Polity Press, Cambridge.

CountryEnergy (2005) Home Energy Efficiency Trial (HEET): Period 2 Survey Results, Country Energy, Sydney.

Cox, S. (2010) Losing Our Cool: Uncomfortable Truths about Our Air-Conditioned World (and Finding New Ways to Get through Summer), The New York Press, New York.

CRA (2005) *Impact Evaluation of the California Statewide Pricing Pilot*, Charles River and Associates (CRA), California.

Darby, S. (2006) The Effectiveness of Feedback on Energy Consumption: a Review for defra of the Literature on Metering, Billing and Direct Displays, Environmental Change Institute, University of Oxford, Oxford.

Darby, S. (2008) 'Energy feedback in buildings: Improving the infrastructure for demand reduction', *Building Research & Information*, vol. 36, no. 5: 499–508.

Darby, S. (2010) 'Smart metering: What potential for householder engagement?', *Building Research & Information*, vol. 38, no. 5: 442–57.

Darnton, A., Verplanken, B., White, P. and Whitmarsh, L. (2011) Habits, Routines and Sustainable Lifestyles: A Summary Report To the Department of Environment, Food and Rural Affairs (DEFRA), AD Research & Analysis, London.

Davidoff, S., Lee, M., Yiu, C., Zimmerman, J. and Dey, A. (2006) 'Principles of smart home control', in P. Dourish and A. Friday (eds), *UbiComp 2006: Ubiquitous Computing*, Springer Berlin/Heidelberg, vol. 4206: 19–34.

Davidson, G. (2008) 'Down the gurgler: Historical influences on Australian domestic water consumption', in P. Troy (ed.), *Troubled Waters: Confronting the Water Crisis in Australia's Cities*, ANU E Press, Canberra, 37–65.

Davison, A. (2001) *Technology and the Contested Meanings of Sustainability*, State University of New York Press, Albany.

Davison, A. (2004) 'Reinhabiting technology: Ends in means and practice of place', *Technology in Society*, vol. 26, 85–97.

Davison, C., Frankel, S. and Smith, G.D. (1992) 'The limits of lifestyle: Re-assessing 'fatalism' in the popular culture of illness prevention', *Social Science & Medicine*, vol. 34, no. 6: 675–85.

De Certeau, M. (1984) *The Practice of Everyday Life*, University of California Press, Berkeley.

de Dear, R.J. and Brager, G.S. (2002) 'Thermal comfort in naturally ventilated buildings: Revisions to ASHRAE Standard 55', *Energy and Buildings*, vol. 34, 549–61.

DECC (2009) *Smarter Grids: The Opportunity. 2050 Roadmap: Discussion Paper*, Department of Energy and Climate Change, London.

DEWHA (2008) *Energy Use in the Australian Residential Sector 1986–2020*, Australian Government: Department of the Environment, Water, Heritage and the Arts (DEWHA), Canberra, Australia.

Dillahunt, T., Mankoff, J., Paulos, E. and Fussell, S. (2009) 'It's not all about "Green": Energy use in low-income communities', paper presented to

Proceedings of the 11th international conference on Ubiquitous computing, Orlando, Florida.
Dobbyn, J. and Thomas, G. (2005) *Seeing the Light: The Impact of Micro-Generation on Our Use of Energy*, The Hub Research Consultants on behalf of the Sustainable Consumption Roundtable, London.
Dourish, P. and Bell, G. (2011) *Divining a Digital Future*, The MIT Press, Cambridge, Massachusetts.
EES (2006) *Status of Air Conditioners in Australia — Updated with 2005 data*, Energy Efficient Strategies prepared for the National Appliance and Equipment Energy Efficiency Committee (NAEEEC), Canberra, ACT.
Ellul, J. (1976) *The Technological Society*, Alfred A. Knopf, New York.
eMeter (2010) *PowerCentsDC™ Program Final Report*, eMeter Strategic Consulting for the Smart Meter Pilot Program, Inc., Foster City, California.
England, M. (2012) 'Home energy management: Make it relevant!', SmartGridNews.com, 22 March 2012, http://www.smartgridnews.com/artman/publish/Business_Customer_Care/Home-energy-management-Make-it-relevant-4595.html.
EST (2007) The Ampere Strikes Back: How Consumer Electronics Are Taking Over the World, Energy Savings Trust (EST), London.
ETSA (2007) 'ETSA Utilities to expand direct load control project', ETSA Utilities, viewed 7 November 2007, www.etsautilities.com.au/media_release.jsp?xcid=1277.
European Commission (2006) *European SmartGrids Technology Platform: Vision and Strategy for Europe's Electricity Networks of the Future*, Office for Offical Publications of the European Communities, Luxembourg, Belgium.
Evans, R. and Marvin, S. (2006) 'Researching the sustainable city: Three modes of interdisciplinarity', *Environment and Planning A*, vol. 38, no. 6: 1009–28.
Everts, J., Lahr-Kurten and Watson, M. (2011) 'Practice matters! Geographical inquiry and theories of practice', *Erdkunde*, vol. 65, no. 4: 323–34.
Faruqui, A. (2012) 'Chapter 3 – The ethics of dynamic pricing', in F.P. Sioshansi (ed.), *Smart Grid*, Academic Press, Boston, 61–83.
Faruqui, A. and George, S. (2005) 'Quantifying customer response to dynamic pricing', *The Electricity Journal*, vol. 18, no. 4: 53–63.
Faruqui, A., Harris, D. and Hledik, R. (2010) 'Unlocking the €53 billion savings from smart meters int he EU: How increasing the adoption of dynamic tariffs could make or break the EU's smart grid investment', *Energy Policy*, vol. 38, 6222–31.
Faruqui, A., Hledik, R., Newell, S. and Pfeifenberger, H. (2007) 'The power of 5 percent', *The Electricity Journal*, vol. 20, no. 8: 68–77.
Faruqui, A., Hledik, R. and Tsoukalis, J. (2009a) 'The power of dynamic pricing', *The Electricity Journal*, vol. 22, no. 3: 42–56.
Faruqui, A. and Palmer, J. (2011) 'Dynamic pricing and its discontents', *Regulation*, vol. 16, no. Fall: 16–22.

Faruqui, A. and Sergici, S. (2010) 'Household response to dynamic pricing of electricity: A survey of 15 experiments', *Journal of Regulatory Economics*, vol. 38, no. 2: 193–225.

Faruqui, A., Sergici, S. and Sharif, A. (2009b) 'The impact of information feedback on energy consumption – A survey of the experimental evidence', *Energy*, vol. 35, 1598–608.

Fine, G.A. (1990) 'Organizational time: Temporal demands and the experience of work in restaurant kitchens', *Social Forces*, vol. 69, no. 1: 95–114.

Fischer, C. (2008) 'Feedback on household electricity consumption: A tool for saving energy?', *Energy Efficiency*, vol. 1, no. 1: 79–104.

Forty, A. (1986) Objects of Desire: Design and Society 1750–1980, Thames and Hudson, London.

Foucault, M. (1995) *Discipline and Punish: The Birth of the Prison*, Vintage Books, New York.

Fox-Penner, P. (2010) Climate Change, the Smart Grid, and the Future the Electric Utilities, Island Press, Washington.

Fox, J. and Gohn, B. (2011) *Executive Summary: Home Energy Management*, Pike Research, Boulder.

Frew, W. (2006) 'Cool customers get $70 a year from the hot ones', *Sydney Morning Herald*, viewed 28 January 2006, http://www.smh.com.au

Geels, F.W. and Smit, W.A. (2000) 'Failed technology futures: Pitfalls and lessons from a historical survey', *Futures*, vol. 32, no. 9–10: 867–85.

Giddens, A. (1984) The Constitution of Society: Outline of the Theory of Structuration, Polity Press, Cambridge.

Goldman, C.A., Barbose, G.L. and Eto, J.H. (2002) 'California customer load reductions during the electricity crisis: Did they help to keep the lights on?', *Journal of Industry, Competition and Trade*, vol. 2, no. 1: 113–42.

Graham, S. and Thrift, N. (2007) 'Out of order', *Theory, Culture & Society*, vol. 24, no. 3: 1–25.

Gram-Hanssen, K. (2007) 'Teenage consumption of cleanliness: How to make it sustainable?', *Sustainability: Science, Practice & Policy*, vol. 3, no. 2: 1–9.

Gram-Hanssen, K. (2008) 'Consuming technologies—Developing routines', *Journal of Cleaner Production*, vol. 16, 1181–9.

Gram-Hanssen, K. (2009) 'Standby consumption in households analyzed with a practice theory approach', *Research and Analysis*, vol. 14, no. 1: 150–65.

Gram-Hanssen, K. (2010) 'Residential heat comfort practices: Understanding users', *Building Research and Information*, vol. 38, no. 2: 175–86.

Gram-Hanssen, K. (2011) 'Understanding change and continuity in residential energy consumption', *Journal of Consumer Culture*, vol. 11, no. 1: 61–78.

Guy, S. and Marvin, S. (1996) 'Transforming urban infrastructure provision—The emerging logic of demand side management', *Policy Studies*, vol. 17, no. 2: 137–47.

Guy, S., Marvin, S. and Moss, T. (eds) (2011) Shaping Urban Infrastructures: Intermediaries and the Governance of Socio-Technical Networks, Earthscan, London.

Hackett, B. and Lutzenhiser, L. (1985) 'The unity of self and object', *Western Folklore*, vol. 44, no. 4: 317–24.
Hackett, B. and Lutzenhiser, L. (1991) 'Social structures and economic conduct: Interpreting variations in household energy consumption', *Sociological Forum*, vol. 6, no. 3: 449–70.
Hacking, I. (1982) 'Biopower and the avalanch of printed numbers', *Humanities in Society*, vol. 5, 279–95.
Hacking, I. (1990) *The Taming of Chance*, Cambridge University Press, Cambridge.
Hacking, I. (1996) 'Normal people', in D.R. Olsen and N. Torrance (eds), *Modes of Thought: Explorations in Culture and Cognition*, Cambridge University Press, New York, 59–71.
Halkier, B. and Jensen, I. (2011) 'Methodological challenges in using practice theory in consumption research. Examples from a study on handling nutritional contestations of food consumption', *Journal of Consumer Culture*, vol. 11, no. 1: 101–23.
Halkier, B., Katz-Gerro, T. and Martens, L. (2011) 'Applying practice theory to the study of consumption: Theoretical and methodological considerations', *Journal of Consumer Culture*, vol. 11, no. 1: 3–13.
Hallnas, L. and Redstrom, J. (2001) 'Slow technology – Designing for reflection', *Personal Ubiquitous Comput.*, vol. 5, no. 3: 201–12.
Hamilton, B., Thomas, C., Park, S.J. and Choi, J-G. (2012) 'Chapter 16 – The customer side of the meter', in F.P. Sioshansi (ed.), *Smart Grid*, Academic Press, Boston, 397–418.
Han, J. and Kamber, M. (2006) *Data Mining: Concepts and Techniques*, Elsevier, San Francisco.
Hand, M. and Shove, E. (2007) 'Condensing practices: Ways of living with a freezer', *Journal of Consumer Culture*, vol. 7, no. 1: 79–104.
Hand, M., Shove, E. and Southerton, D. (2005) 'Explaining showering: A discussion of the material, conventional, and temporal dimensions of practice', *Sociological Research Online*, vol. 10, no. 2.
Haraway, D. (1991) *Simians, Cyborgs, and Women: The Reinvention of Nature*, Routledge, New York.
Hargreaves, T. (2010) Working paper 141: The visible energy trial: Insights from qualitiative interviews, Tyndall Centre for Climate Change Research, Norwich.
Hargreaves, T., Nye, M. and Burgess, J. (2010) 'Making energy visible: A qualitative field study of how householders interact with feedback from smart energy monitors', *Energy Policy*, vol. 38, 6111–9.
Hargreaves, T., Nye, M. and Burgess, J. (2013) 'Keeping energy visible? Exploring how householders interact with feedback from smart energy monitors in the longer term', *Energy Policy*, vol. 52, 126–34.
Harper-Slaboszewicz, P., McGregor, T. and Sunderhauf, S. (2012) 'Chapter 15 – Customer view of smart grid—set and forget?', in F.P. Sioshansi (ed.), *Smart Grid*, Academic Press, Boston, 371–95.

Harris (2012) 'Americans making changes to be more energy efficient at home', Harris Interactive: Harris Polls, 22 March 2012, http://www.harrisinteractive.com/NewsRoom/HarrisPolls/tabid/447/mid/1508/articleId/980/ctl/ReadCustom%20Default/Default.aspx.

Hauber-Davidson, G. and Idris, E. (2006) 'Smart water metering', *Water*, vol. 33, no.3: 38–41.

Hawkins, G. and Race, K. (2011) 'Bottle water practices: Reconfiguring drinking in Bangkok households', in R. Lane and A. Gorman-Murray (eds), *Material Geographies of Household Sustainability*, Ashgate, Farnham, 113–24.

Head, L. (2008) 'Nature, networks and desire: Changing cultures of water in Australia', in P. Troy (ed.), *Troubled Waters: Confronting the Water Crisis in Australia's Cities*, ANU E Press, Canberra, ACT, 67–80.

Healy, S. and MacGill, I. (2012) 'Chapter 2 – From smart grid to smart energy use', in F.P. Sioshansi (ed.), *Smart Grid*, Academic Press, Boston, 29–59.

Herter, K. (2007) 'Residential implementation of critical-peak pricing of electricity', *Energy Policy*, vol. 35, 2121–30.

Herter, K., McAuliffe, P. and Rosenfeld, A. (2007) 'An exploratory analysis of California residential customer response to critical peak pricing of electricity', *Energy*, vol. 32, 25–34.

Heschong, L. (1979) *Thermal Delight in Architecture*, Masachusetts Institute of Technology, Cambridge, Massachusetts.

Hillier, J. and Rooksby, E. (2005) 'Introduction to first edition', in J. Hillier and E. Rooksby (eds), *Habitus: A Sense of Place*, 2nd edn, Ashgate Publishing Ltd, Hants, 19–25.

Hinchliffe, S. (1996) 'Helping the earth begins at home: The social construction of socio-environmental responsibilities', *Global Environmental Change*, vol. 6, no. 1: 53–62.

Hitchings, R. (2007) 'Geographies of embodied outdoor experience and the arrival of the patio heater', *Area*, vol. 39, no. 3: 340–8.

Hitchings, R. (2012) 'People can talk about their practices', *Area*, vol. 44, no. 1: 61–7.

Hledik, R. (2009) 'How green is the smart grid?', *The Electricity Journal*, vol. 22, no. 3: 29–41.

Hobson, K. (2011) 'Environmental politics, green governmentality and the possibility of a 'creative grammar' for domestic sustainable consumption', in R. Lane and A. Gorman-Murray (eds), *Material Geographies of Household Sustainability*, Ashgate, Farnham, 193–210.

Honebein, P.C., Cammarano, R.F. and Donnelly, K.A. (2009) 'Will smart meters ripen or rot? Five first principles for embracing customers as co-creators of value', *The Electricity Journal*, vol. 22, no. 5: 39–44.

Horrigan, B. (1986) 'The home of tomorrow, 1927–1945', in J.J. Corn (ed.), *Imagining Tomorrow: History, Technology and the American Future*, The MIT Press, Cambridge, Massachusetts, 137–63.

House of Commons (2007) *Local Energy: Turning Consumers into Producers*, House of Commons, Trade and Industry Committee, London.

Hughes, T. P. (1983) *Networks of Power: Electrification in Western Society*, Johns Hopkins University Press, Baltimore.

Humphreys, M.A. and Nicol, J.F. (1998) 'Understanding the adaptive approach to thermal comfort', in ASHRAE Technical Data Bulletin, American Society of Heating, Refrigeration and Air-Conditioning Engineers (ASHRAE): 1–14.

IBM (2011) 'IBM survey reveals new type of energy concern: Lack of consumer understanding', *IBM News Room*, viewed 7 March 2012, http://www-03.ibm.com/press/us/en/pressrelease/35271.wss.

IEA (2005) *Saving Electricity in a Hurry: Dealing with Temporary Shortfalls in Electricity Supplies*, Organisation for Economic Cooperation and Development (OECD) and International Energy Agency (IEA), Paris.

Intille, S.S. (2002) 'Designing a home of the future', *Pervasive Computing*, vol. 1, no. 2: 76–82.

Ipsos MORI (2012) 'Ofgem consumer first panel Year 4: Findings from first workshops (held October and November 2011)', Ipsos MORI for the Office of Gas and Electricity Markets (Ofgem), London.

Jalas, M. and Rinkinen, J. (2012) 'Stacking wood and staying warm: Sequence, rhythm, synchronization and tempos of domestic wood-based heating practices', paper presented to EASST and 4S, Copenhagen, October.

Jelly, R. (2008) *Phase 2: Qualitative Assessment of Consumer Responses to the National Electricity Smart Meter Rollout Program*, Red Jelly, prepared for NERA Economic Consulting on behalf of the Department of Industry, Science and Tourism, Sydney.

Juntunen, J.K. (2011) 'Domestication pathways of small-scale renewable energy technologies', in M. Matsumoto, Y. Umeda, K. Masui and S. Fukushige (eds), Conference proceedings for EcoDesign2011: 7th International Symposium on Environmentally Conscious Design and Inverse Manufacturing, SpringerLink, Kyoto: 164–9.

Kaika, M. (2005) City of Flows: Modernity, Nature, and the City, Routledge, New York.

Kaufmann, C. (1998) *Dirty Linen: Couples and their Laundry*, Middlesex University Press, London.

Keirstead, J. (2007) 'Behavioural responses to photovoltaic systems in the UK domestic sector', *Energy Policy*, vol. 35, no. 8: 4128–41.

Kempton, W., Feuermann, D. and McGarity, A.E. (1992a) '"I always turn it on super": User decisions about when and how to operate room air conditioners', *Energy and Buildings*, vol. 18, 177–91.

Kempton, W. and Layne, L.L. (1994) 'The consumer's energy analysis environment', *Energy Policy*, vol. 22, no. 10: 857–66.

Kempton, W. and Montgomery, L. (1982) 'Folk quantification of energy', *Energy*, vol. 7, no. 10: 817–27.

Kempton, W., Reynolds, C., Fels, M. and Hull, D. (1992b) 'Utility control of residential cooling: Resident-perceived effects and potential program improvements', *Energy and Buildings*, vol. 18, 201–19.

King, C. and Delurey, D. (2005) 'Efficiency and demand response: Twins, siblings, or cousins?', *Public Utilities Fortnightly*, March: 54–61

Kling, R. (1994) 'Reading "all about" computerization: How genre conventions shape nonfiction social analysis', *The Information Society*, vol. 10, no. 3: 147–72.

Klopfert, F. and Wallenborn, G. (2011) *Empowering Consumers through Smart Metering*, Bureau Eruopeen des Unions des Consommateurs (BEUC), Brussels.

Koskela, T., Väänänen-Vainio-Mattila, K. (2004) 'Evolution towards smart home environments: Empirical evaluation of three user interfaces', *Personal Ubiquitous Computing*, vol. 8, no. 3–4: 234–40.

Kuijer, L. and De Jong, A. (2011) 'Practice theory and human centred design: A sustainable bathing example', in Nordic Design Research Conference, Helsinki.

Kuijer, L. and De Jong, A. (2012) 'Identifying design opportunities for reduced household resource consumption: Exploring practices of thermal comfort', *Journal of Design Research*, vol. 10, no. 1: 67–85.

Kurz, T., Donaghue, N., Rapley, M. and Walker, I. (2005) 'The ways that people talk about natural resources: Discursive strategies as barriers to environmentally sustainable practices', *British Journal of Social Psychology*, vol. 44, 603–20.

Latour, B. (1987a) Science in Action: How to Follow Scientists and Engineers through Society, Harvard University Press, Cambridge, Massachusetts.

Latour, B. (1987b) 'Where are the missing masses? The sociology of a few mundane artifacts', in W.E. Bijker, T. Hughes and T. Pinch (eds), *The Social Construction of Technological Systems: New Directions in the Sociology and History of Technology*, The MIT Press, Cambridge, Massachusetts, 225–58.

Latour, B. (2000) 'When things strike back: A possible contribution of "science studies" to the social sciences', *The British Journal of Sociology*, vol. 51, no. 1: 107–23.

Latour, B. (2005) Reassembling the Social: An Introduction to Actor-Network-Theory, Oxford University Press, New York.

Law, J. (1993) *Modernity, Myth and Materialism*, Blackwell, Oxford.

Law, J. (2004) After Method: Mess in Social Science Research, Routledge, New York.

Law, J. (2009) 'Seeing Like a Survey', *Cultural Sociology*, vol. 3, no. 2: 239–56.

Leenheer, J., de Nooij, M. and Sheikh, O. (2011) 'Own power: Motives of having electricity without the energy company', *Energy Policy*, vol. 39, no. 9: 5621–9.

Lefebvre, H. (2004) Rhythmanalysis: Space, Time and Everyday Life, Continuum, London.

Leshed, G. and Sengers, P. (2011) '"I lie to myself that i have freedom in my own schedule": Productivity tools and experiences of busyness', paper presented to Proceedings of the 2011 annual conference on Human factors in computing systems, Vancouver.

Licoppe, C. (2010) 'The "performative turn" in science and technology studies', *Journal of Cultural Economy*, vol. 3, no. 2: 181–8.

Lien, M.E. and Law, J. (2011) '"Emergent Aliens": On salmon, nature, and their enactment', *Ethnos*, vol. 76, no. 1: 65–87.

Lindsay, J. (2010) 'Healthy living guidelines and the disconnect with everyday life', *Critical Public Health*, vol. 20, no. 4: 475–87.

Ling, R. and Yttri, B. (2002) 'Hyper-coordination via mobile phones in Norway', in J.E. Katz and M.A. Aakhus (eds), *Perpetual Contact: Mobile Communication, Private Talk, Public Performance*, Cambridge University Press, Cambridge, 139–69.

Livingstone, S. (1992) 'The meaning of domestic technologies: A personal construct analysis of familial gender relations', in R. Silverston and E. Hirsch (eds), *Consuming Technologies: Media and Information in Domestic Spaces*, Routledge, London, 113–30.

Logan, R.J., Augaitis, S., Miller, R.H. and Wehmeyer, K. (1995) 'Living room culture—An anthropological study of television usage behaviors', *Proceedings of the Human Factors and Ergonomics Society Annual Meeting*, vol. 39, no. 5: 326–30.

Logica (2010) 2010 Australian Smart Grid Study: a Comprehensive View of the Strategies, Priorities, and Challenges for Smart Grid Adoption in Australia, Logica, Sydney.

Lohman, T. (2011) 'Ausgrid spruiks smart grid consumer tools', *Computerworld*, viewed 15 December, http://www.computerworld.com.au/article/410340/ausgrid_spruiks_smart_grid_consumer_tools/.

Lutron (2012) Experience the Essence of Pleasure (Promotional Flyer), Lutron, New York.

Lutzenhiser, L. (1997) 'Social structure, culture, and technology: Modeling the driving forces of household energy consumption', in P.C. Stern, T. Dietz, V.W. Ruttan, R.H. Socolow and J.L. Sweeney (eds), *Environmentally Significant Consumption: Research Directions*, National Academy Press, Washington, 77–91.

Lutzenhiser, L. (2009) *Behavioral Assumptions Underlying California Residential Sector Energy Efficiency Programs*, California Institute for Energy and Environment for CIEE Behavior and Energy Program, Oakland.

Lutzenhiser, L., Gossard, M.H. and Bender, S. (2002) 'Crisis in paradise: Understanding the household conservation response to California's 2001 energy crisis', in Proceedings of the 2002 ACEEE Summer Study, American Council for an Energy-Efficient Economy, Washington.

Macnaghten, P. (2003) 'Embodying the environment in everyday life practices', *The Sociological Review*, vol. 51, no. 1: 63–84.

Maller, C. (2011) 'Practices involving energy and water consumption in migrant households', in P. Newton (ed.), *Urban Consumption*, CSIRO Publishing, Collingwood, 237–50.

Maller, C., Horne, R. and Dalton, T. (2011) 'Green renovations: Intersections of daily routines, housing aspirations and narratives of environmental sustainability', *Housing, Theory and Society*, vol. 29, no. 3: 255–75.

Maller, C. and Strengers, Y. (2013) 'The global migration of everyday life: Investigating the practice memories of Australian migrants', *Geoforum*, vol. 44, 243–52.

Marres, N. (2010) 'Front-staging nonhumans: Publicity as a constraint on the political activity of things', in B. Braun, S. Wahmore and I. Stengers (eds), *Political Matter: Technoscience, Democracy, and Public Life*, University of Minnesota Press, Minneapolis, 177–209.

Marres, N. (2011) 'The costs of public involvement: Everyday devices of carbon accounting and the materialization of participation', *Economy and Society*, vol. 40, no. 4: 510–33.

Marres, N. (2012a) *Material Participation: Technology, the Environment and Everyday Publics*, Palgrave Macmillan, London.

Marres, N. (2012b) 'On some uses and abuses of topology in the social analysis of technology (or the problem with smart meters)', *Theory, Culture & Society*, vol. 29, no. 4–5: 288–310.

Marvin, S., Chappells, H. and Guy, S. (1999) 'Pathways of smart metering development: Shaping environmental innovation', *Computers, Environment and Urban Systems*, vol. 23, 109–26.

Marvin, S., Chappells, H. and Guy, S. (2011) 'Smart meters as obligatory intermediaries: Reconstructing environmental action', in S. Guy, S. Marvin, W. Medd and T. Moss (eds), *Shaping Urban Infrastructures: Intermediaries and the Governance of Socio-Technical Networks*, Earthscan, London, 175–91.

Mayer-Schönberger, V. and Cukier, K. (2013) *Big Data: a Revolution That Will Transform How We Live, Work, and Think*, Houghton Mifflin Harcourt Publishing Company, New York.

McGowan, S. (2009) 'Hot in the city', *Ecolibrium*, vol. February, 14–7.

Meier, A., Aragon, C., Peffer, T. and Pritoni, M. (2010) *Thermostat Interface and Usability: A Survey*, prepared for Ernest Orlando Lawrence Berkeley National Laboratory, Berkeley.

Michael, M. (2006) *Technoscience and Everyday Life*, Open University Press, Berkshire.

Miller, P. and Rose, N. (1997) 'Mobilizing the consumer', *Theory, Culture & Society*, vol. 14, no. 1: 1–36.

More, T. (2005) *Utopia*, Digireads.com Publishing, Stilwell.

Mozer, M.C. (2005) 'Lessens from an adaptive home', in D.J. Cook and S.K. Das (eds), *Smart Environments: Technologies, Protocols and Applications*, John Wiley & Sons, New Jersey, 273–94.

NCES (2012) *Digest of Education Statistics 2011*, National Centre for Education Statistics (NCES), Washington.

NERA (2008a) Cost-Benefit Analysis of Smart Metering and Direct Load Control. Work Stream 4: Consumer Impacts. Phase 2 Consultation Report, NERA Economic Consulting, prepared for the Ministerial Council on Energy Smart Meter Working Group, Sydney.

NERA (2008b) *Cost Benefit Analysis of Smart Metering and Direct Load Control: Overview Report for Consultation*, NERA Economic Consulting, prepared for the Ministerial Council on Energy Smart Meter Working Group, Sydney.

Newsham, G.R., Birt, B.J. and Rowlands, I.H. (2011) 'A comparison of four methods to evaluate the effect of a utility residential air-conditioner load control program on peak electricity use', *Energy Policy*, vol. 39, no. 10: 6376–89.

Newsham, G.R. and Bowker, B.G. (2010) 'The effect of utility time-varying pricing and load control strategies on residential summer peak electricity use: A review', *Energy Policy*, vol. 38, no. 7: 3289–96.

Ngar-yin Mah, D., van der Vleuten, J.M., Hills, P. and Tao, J. (2012) 'Consumer perceptions of smart grid development: Results of a Hong Kong survey and policy implications', *Energy Policy*, vol. 49, 204–16.

Nicol, F. and Roaf, S. (2007) 'Progress on passive cooling: Adaptive thermal comfort and passive architecture', in M. Santamouris (ed.), *Advances in Passive Cooling*, Earthscan, London, 1–29.

Nye, D.E. (2010) *When the Lights Went Out: a History of Blackouts in America*, The MIT Press, Cambridge, Massachusetts.

OETD (2003) *"Grid 2030" A national vision for electricity's second 100 years*, United States Department of Energy, Office of Electric Transmission and Distribution (OETD), USA.

Oksanen-Sarela, K. and Pantzar, M. (2001) 'Smart life, version 3.0: Representations of everyday life in future studies', in J. Gronow and A. Warde (eds), *Ordinary Consumption*, Routledge, London, 199–218.

OL (2009) *Ofgem Consumer First Panel: The 'Future Consumer'*, Opinion Leader (OL) for the Office of Gas and Electricity Markets, London.

OSTP (2012) 'New industry commitments to give 15 million households tools to shrink their energy bills', Office of Science and Technology Policy (OSTP), Washington, 26 March 2012, http://www.whitehouse.gov/administration/eop/ostp/pressroom/03222012.

Pamlin, D. (2002) 'A report about ICT and sustainability', in D. Pamlin (ed.), *Sustainability at the Speed of Light: Opportunities and Challenges for Tomorrow's Society*, WWF Sweden, Stockholm, 5–10.

Pantzar, M. and Shove, E. (2010a) 'Temporal rhythms as outcomes of social practices: A speculative discussion', *Ethnologia Europaea*, vol. 40, no. 1: 19–29.

Pantzar, M. and Shove, E. (2010b) 'Understanding innovation in practice: A discussion of the production and reproduction of Nordic Walking', *Technology Analysis & Strategic Management*, vol. 22, no. 4: 447–61.

PCE (2009) *Smart Electricity Meters: How Households and the Environment can Benefit*, Parliamentary Commissioner for the Environment (PCE), Wellington.

Petersen, A. and Lupton, D. (1996) *The New Public Health: Health and Self in the Age of Risk*, Allen & Unwin, St Leonards.

PG&E (2012) 'Take control with SmartMeter data', Pacific Gas and Electricity Company (PG&E), 21 March 2012, http://www.pge.com/smartmeter/customerstories/control.shtml.

Pickering, A. (1993) 'The mangle of practice: Agency and emergence in the sociology of science', *American Journal of Sociology*, vol. 99, no. 3: 559–89.

Pickering, A. (1995) *The Mangle of Practice: Time, Agency and Science*, University of Chicago Press, Chicago.

Pierce, J., Fan, C., Lomas, D., Marcu, G. and Paulos, E. (2010a) 'Some considerations of the (in)effectiveness of residential energy feedback systems', paper presented to Designing Interactive Systems (DIS) 2010, Aarhus, 16–20 August.

Pierce, J., Schiano, D.J. and Paulos, E. (2010b) 'Home, habits, and energy: Examining domestic interactions and energy consumption', in CHI 2010, ACM, Atlanta, 10–15 April 1985–94.

Pierce, J., Brynjarsdottir, H., Sengers, P. and Strengers, Y. (2011) 'Everyday practice and sustainable HCI: Understanding and learning from cultures of (un)sustainability', paper presented to CHI '11 Extended Abstracts on Human Factors in Computing Systems, Vancouver.

Pierce, J. and Paulos, E. (2010) 'Materializing energy', paper presented to Proceedings of the 8th ACM Conference on Designing Interactive Systems, Aarhus.

Pierce, J. and Paulos, E. (2012a) 'Beyond energy monitors: Interaction, energy, and emerging energy systems', paper presented to Proceedings of the 2012 ACM annual conference on Human Factors in Computing Systems, Austin.

Pierce, J. and Paulos, E. (2012b) 'Designing everyday technologies with human-power and interactive microgeneration', paper presented to Proceedings of the Designing Interactive Systems Conference, Newcastle Upon Tyne.

Pierce, J. and Paulos, E. (2012c) 'The local energy indicator: Designing for wind and solar energy systems in the home', paper presented to Proceedings of the Designing Interactive Systems Conference, Newcastle Upon Tyne.

Pike (2011) 'Home Energy Management Users to Reach 63 Million by 2020', Pike Research, 23 March 2012, http://www.pikeresearch.com/newsroom/home-energy-management-users-to-reach-63-million-by-2020>.

Pink, S. (2004) Home Truths: Gender, Domestic Objects and Everyday Life, Berg, Oxford.

Pink, S. (2005) 'Dirty laundry. Everyday practice, sensory engagement and the constitution of identity', Social Anthropology, vol. 13, no. 3: 275–90.

Pink, S. (2007) 'Walking with video', Visual Studies, vol. 22, no. 3: 240–52.

Pink, S. (2012a) 'Ethnography of the invisible: Energy in the multisensory home', Ethnologia Europaea, vol. 41, no. 1: 117–28.

Pink, S. (2012b) Situating Everyday Life, SAGE, London.

Pink, S., Tutt, D., Dainty, A. and Gibb, A. (2010) 'Ethnographic methodologies for construction research: Knowing, practice and interventions', Building Research & Information, vol. 36, no. 6: 647–59.

Platt, G., Berry, A. and Cornforth, D. (2012) 'Chapter 8 – What role for microgrids?', in F.P. Sioshansi (ed.), Smart Grid, Academic Press, Boston, 185–207.

Polak, F. (1973) The Image of the Future, Jossey-Bass Inc., San Francisco.

Postill, J. (2010) 'Introduction: Theorising media and practice', in B. Bräuchler and J. Postill (eds), Theorising Media and Practice, Berghahn Books, New York, 1–32.

Prins, G. (1992) 'On condis and coolth', Energy and Buildings, vol. 18, 251–8.

Pullinger, M., Browne, A., Anderson, B. and Medd, W. (2013) Patterns of Water: The Water Related Practices of Households in Southern England, and Their Influence on Water Consumption and Demand Management, Lancaster University, Lancaster.

Purpura, S., Schwanda, V., Williams, K., Stubler, W. and Sengers, P. (2011) 'Fit4life: The design of a persuasive technology promoting healthy behavior and ideal weight', paper presented to Proceedings of the 2011 annual conference on Human factors in computing systems, Vancouver.

Ramchurn, S.D., Vytelingum, P., Rogers, A. and Jennings, N.R. (2012) 'Puting the "smarts" into the smart grid: A grand challenge for artificial intelligence', *Communications of the ACM*, vol. 55, no. 4, April: 86–97.
Reckwitz, A. (2002a) 'The status of the "material" in theories of culture. From "social structure" to "artefacts"', *Journal for the Theory of Social Behaviour*, vol. 32, no. 2: 195–217.
Reckwitz, A. (2002b) 'Toward a theory of social practices: A development in culturalist theorizing', *Journal of Social Theory*, vol. 5, no. 2: 243–63.
Reeves, B., Cummings, J.J., Scarborough, J.K., Flora, J. and Anderson, D. (2012) 'Leveraging the engagement of games to change energy behavior', in Collaboration Technologies and Systems (CTS), 2012 International Conference on, 21–25 May 2012: 354–8.
Reidy, C. (2006) *Interval Meter Technology Trials and Pricing Experiments: Issues for Small Consumers*, Institute of Sustainable Futures, prepared for Consumer Utilities Advocacy Centre, Sydney.
Reidy, C., Wilson, E., Cheney, H. and Tarlo, K. (2005) *Community EmPOWERment Research Report Summary*, Institute for Sustainable Futures, Melbourne.
RMI (2006) *Demand Response: An Introduction. Overview of Programs, Technologies, and Lessons Learned*, Rocky Mountain Institute, prepared for the Southwest Energy Efficiency Project, Boulder.
Rode, J.A., Toye, E.F. and Blackwell, A.F. (2004) 'The fuzzy felt ethnography—Understanding the programming patterns of domestic appliances', *Personal Ubiquitous Comput.*, vol. 8, no. 3–4: 161–76.
Røpke, I. (2009) 'Theories of practice – New inspiration for ecological economic studies of consumption', *Ecological Economics*, vol. 68: 2490–7.
Røpke, I. and Christensen, T.H. (2012) 'Energy impacts of ICT – Insights from an everyday life perspective', *Telematics and Informatics*, vol. 29, no. 4: 348–61.
Røpke, I., Haunstrup Christensen, T. and Ole Jensen, J. (2010) 'Information and communication technologies – A new round of household electrification', *Energy Policy*, vol. 38, no. 4: 1764–73.
Rose, N. and Miller, P. (2010) 'Political power beyond the state: Problematics of government', *The British Journal of Sociology*, vol. 61: 271–303.
Roy, R., Caird, S. and Abelman, J. (2008) YIMBY Generation – Yes in My Back Yard! uk Householders Pioneering Microgeneration Heat, The Energy Saving Trust, London.
Sauter, R. and Watson, J. (2007) 'Strategies for the deployment of microgeneration: Implications for social acceptance', *Energy Policy*, vol. 35, no. 5: 2770–9.
Schatzki, T. (2010) 'Materiality and social life', *Nature and Culture*, vol. 5, no. 2: 123–49.
Schatzki, T.R. (1996) Social Practices: A Wittgensteinian Approach to Human Activity and the Social, Cambridge University Press, Cambridge.
Schatzki, T.R. (1997) 'Practices and actions: A Wittgensteinian critique of Bourdieu and Giddens', *Philosophy of the Social Sciences*, vol. 27, no. 3: 283–308.

Schatzki, T.R. (2001) 'Introduction: Practice theory', in T.R. Schatzki, K. Knorr Cetina and E. Von Savigny (eds), *The Practice Turn in Contemporary Theory*, Routledge, New York, 1–14.
Schatzki, T.R. (2002) *The Site of the Social: A Philosophical Account of the Constitution of Social Life and Change*, The Pennsylvania State University Press, Pennsylvania, USA.
Schatzki, T.R., Knorr Cetina, K. and Von Savigny, E. (2001) *The Practice Turn in Contemporary Theory*, Routledge, New York.
Schleicher-Tappeser (2012) *The Smart Grids Debate in Europe*, Smart Energy for Europe Platform, Berlin.
Scholem, G. (1965) *On the Kabbalah and Its Symbolism*, Schocken Books, New York.
Scholem, G. (1966) 'The Golem of Prague and the Golem of Rehovoth', *Observations*: 62–6
Schumacher, E.F. (1999) Small Is Beautiful: Economics as If People Mattered: 25 Years Later...with Commentaries, Hartley & Marks Publishers, Vancouver.
Schwartz Cowan, R. (1989) More Work for Mother: The Ironies of Household Technology from the Open Hearth to the Microwave, Free Association Books, London.
Scott, K., Bakker, C. and Quist, J. (2012) 'Designing change by living change', *Design Studies*, vol. 33, no. 3: 279–97.
Segal, H. (1986) 'The technological utopians', in J.J. Corn (ed.), *Imagining Tomorrow: History, Technology and the American Future*, The MIT Press, Cambridge, Massachusetts, 119–36.
Seligman, C., Darley, J.M. and Becker, L.J. (1978) 'Behavioral approaches to residential energy conservation', in R.H. Socolow (ed.), *Saving Energy in the Home: Princeton's Experiments at Twin Rivers*, Ballinger Publishing Company, Massachusetts, 231–54.
SGA (2011) *Maximising Consumer Benefits*, Smart Grid Australia (SGA), Melbourne.
SGCC (2012) *2012 State of the Consumer Report*, Smart Grid Consumer Collaborative, Roswell.
Shove, E. (1997) 'Revealing the invisible: Sociology, energy and the environment', in M. Redclift and G. Woodgate (eds), *The International Handbook of Environmental Sociology*, Edward Elgar Publishing, Cheltenham, 271–3.
Shove, E. (2003) Comfort, Cleanliness and Convenience: The Social Organisation of Normality, Berg Publishers, Oxford.
Shove, E. (2004) 'Efficiency and consumption: Technology and practice', *Energy & Environment*, vol. 15, no. 6: 1053–65.
Shove, E. (2010a) 'Beyond the ABC: Climate change policy and theories of social change', *Environment and Planning A*, vol. 42: 1273–85.
Shove, E. (2010b) 'Social theory and climate change: Questions often, sometimes and not yet asked', *Theory, Culture & Society*, vol. 27, no. 2–3: 277–88.
Shove, E. (2011) 'Commentary: On the different between chalk and cheese – A response to Whitmarsh et al.'s comments on "Beyond the ABC: Climate

change policy and theories of social change"', *Environment and Planning A*, vol. 43: 262–4.

Shove, E. & Chappells, H. (2001) 'Ordinary consumption and extraordinary relationships: Utilities and their users', in J. Gronow and A. Warde (eds), *Ordinary Consumption*, Routledge, London, 45–58.

Shove, E., Lutzenhiser, L., Guy, S., Hackett, B. & Wilhite, H. (1998) 'Energy and social systems', in S. Rayner and E.L. Malone (eds), *Human Choice and Climate Change*, Batelle Press, Ohio, 291–323.

Shove, E. and Pantzar, M. (2005a) 'Consumers, producers and practices: Understanding the invention and reinvention of Nordic walking', *Journal of Consumer Culture*, vol. 5, no. 1: 43–64.

Shove, E. and Pantzar, M. (2005b) 'Fossilisation', *Ethnologia Europaea*, vol. 35: 59–63.

Shove, E. and Pantzar, M. (2007) 'Recruitment and reproduction: The carriers of digital photography and floorball', *Human Affairs*, vol. 17: 154–67.

Shove, E., Pantzar, M. and Watson, M. (2012) The Dynamics of Social Practice: Everyday Life and How it Changes, SAGE, London.

Shove, E., Trentmann, F. and Wilk, R. (eds) (2009) Time, Consumption and Everyday Life: Practice, Materiality and Culture, Berg, Oxford.

Shove, E. and Walker, G. (2010) 'Governing transitions in the sustainability of everyday life', *Research Policy*, vol. 39: 471–6.

Shove, E., Watson, M., Hand, M. and Ingram, J. (2007) *The Design of Everyday Life*, Berg, Oxford.

Sioshansi, F.P. (2012) 'Introduction', in F.P. Sioshansi (ed.), *Smart Grid*, Academic Press, Boston, xxix–lvi.

Slob, A. and Verbeek, P.-P. (2006) 'Technology and user behavior: An introduction', in P.-P. Verbeek and A. Slob (eds), *User Behavior and Technology Development: Shaping Sustainable Relations between Consumers and Technologies*, Springer, Dordrecht.

Sofoulis, Z. (2005) 'Big water, everyday water: A sociotechnical perspective', *Continuum: Journal of Media & Cultural Studies*, vol. 19, no. 4: 445–63.

Sofoulis, Z. (2011) 'Skirting complexity: The retarding quest for the average water user', *Continuum: Journal of Media & Cultural Studies*, vol. 25, no. 6: 795–810.

Southerton, D. (2003) 'Squeezing time': Allocating practices, coordinating networks and scheduling society', *Time & Society*, vol. 12, no. 1: 5–25.

Southerton, D. (2006) 'Analysing the temporal organization of daily life: Social constraints, practices and their allocation', *Sociology*, vol. 40, no. 3: 435–54.

Southerton, D. (2007) 'Time pressure, technology and gender: The conditioning of temporal experiences in the UK', *Equal Opportunities International*, vol. 26, no. 2: 113–28.

Southerton, D. (2009) 'Re-ordering Temporal Rhythms: Coordinating daily practices in the UK in 1937 and 2000', in E. Shove, F. Trentmann and R.R. Wilk (eds), *Time, Consumption and Everyday Life: Practice, Materiality and Culture*, Berg, Oxford, 49–63.

Southerton, D., Chappells, H. & Van Vliet, B. (eds) (2004) *Sustainable Consumption: The Implications of Changing Infrastructures of Provision*, Edward Elgar, Cheltenham.

Spaargaren, G. (2011) 'Theories of practices: Agency, technology, and culture: Exploring the relevance of practice theories for the governance of sustainable consumption practices in the new world-order', *Global Environmental Change*, vol. 21, no. 3: 813–22.

Star, S.L. (1999) 'The ethnography of infrastructure', *American Behavioral Scientist*, vol. 43, no. 3: 377–91.

Stern, P. (1986) 'Blind spots in policy analysis: What economics doesn't say about energy use', *Journal of Policy Analysis and Management*, vol. 5, no. 2: 200–27.

Strengers, Y. (2008) 'Comfort expectations: The impact of demand-management strategies in Australia', *Building Research and Information*, vol. 36, no. 4: 381–91.

Strengers, Y. (2009) 'Bridging the divide between resource management and everyday life: Smart metering, comfort and cleanliness', *PhD Thesis*, RMIT University.

Strengers, Y. (2010) 'Air-conditioning Australian households: A trial of Dynamic Peak Pricing', *Energy Policy*, vol. 38, no. 11: 7312–22.

Strengers, Y. (2011a) 'Beyond demand management: Co-managing energy and water consumption in Australian households', *Policy Studies*, vol. 32, no. 1: 35–58.

Strengers, Y. (2011b), 'Designing eco-feedback systems for everyday life', in *Proceedings of the 2011 Annual Conference on Human Factors in Computing Systems*, ACM, Vancouver, 2135–44.

Strengers, Y. (2011c) 'Negotiating everyday life: The role of energy and water consumption feedback', *Journal of Consumer Culture*, vol. 11, no. 19: 319–38.

Strengers, Y. (2012a) 'Peak electricity demand and social practice theories: Reframing the role of change agents in the energy sector', *Energy Policy*, vol. 44: 226–34.

Strengers, Y. (2012b) 'Interdisciplinarity and industry collaboration in doctoral candidature: Tensions within and between discourses', *Studies in Higher Education*, 1–14, DOI: 10.1080/03075079.2012.709498.

Strengers, Y. and Maller, C. (2011) 'Integrating health, housing and energy policies: The social practices of cooling', *Building Research & Information*, vol. 39, no. 2: 154–68.

Strengers, Y. and Maller, C. (2012) 'Materialising energy and water resources in everyday practices: Insights for securing supply systems', *Global Environmental Change*, vol. 22, no. 3: 754–63.

Strother, N. and Gohn, B. (2012) *Executive Summary: Home Energy Management*, Pike Research, Boulder.

Sturken, M. and Thomas, D. (2004) 'Introduction: Technological visions and the rhetoric of the new', in M. Struken, D. Thomas and S. Ball-Rokeach (eds), *Technological Visions: The Hopes and Fears that Shape New Technologies*, Temple University Press, Philadephia, 1–18.

Taylor, V., Chappells, H., Medd, W. and Trentmann, F. (2009) 'Drought is normal: The socio-technical evolution of drought and water demand in England and Wales, 1893–2006', *Journal of Historical Geography*, vol. 35, no. 3: 568–91.

Thaler, R.H. and Sunstein, C.T. (2008) *Nudge: Improving Decisions about Health, Wealth, and Happiness*, Yale University Press, New Haven.

Toffler, A. (1980) *The Third Wave*, Bantam, New York.

Trentmann, F. (2006) 'Knowing consumers – Histories, identities, practices: An introduction', in F. Trentmann (ed.), *The Making of the Consumer: Knowledge, Power and Identify in the Modern World*, Berg, Oxford, 1–29.

Trentmann, F. (2009) 'Disruption is normal: Blackouts, breakdowns and the elasticity of everyday life', in E. Shove, F. Trentmann and R.R. Wilk (eds), *Time, Consumption and Everyday Life: Practice, Materiality and Culture*, Berg, Oxford, 67–84.

Truninger, M. (2011) 'Cooking with Bimby in a moment of recruitment: Exploring conventions and practice perspectives', *Journal of Consumer Culture*, vol. 11, no. 37: 37–59.

Valocchi, M. and Juliano, J. (2012) *Knowledge is Power: Driving Smarter Energy Usage through Consumer Education*, IBM Institute for Business Value, Somers.

Valocchi, M., Juliano, J. and Schurr, A. (2009) *Lighting the Way: Understanding the Smart Energy Consumer*, IBM Institute for Business Value, Somers.

Valocchi, M., Schurr, A., Juliano, J. and Nelson, E. (2007) *Plugging in the Consumer: Innovating Utility Business Models for the Future*, IBM Institute for Business Value, Somers.

Van Vliet, B. (2006) 'Citizen-consumer roles in environment management of large technological systems', in P.-P. Verbeek and A. Slob (eds), *User Behavior and Technology Development: Shaping Sustainable Relations between Consumers and Technologies*, Springer, Dordrecht, 309–18.

Van Vliet, B., Chappells, H. and Shove, E. (2005) Infrastructures of Consumption: Environmental Innovation in the Utilities Industries, Earthscan, London.

Vassileva, I., Odlare, M., Wallin, F. and Dahlquist, E. (2012) 'The impact of consumers' feedback preferences on domestic electricity consumption', *Applied Energy*, vol. 93: 575–82.

Velthuis, O. (2004) 'An interpretive approach to meanings of prices', *The Review of Austrian Economics*, vol. 17, no. 4: 371–86.

Vermeer, T. (2008) 'Big Brother plan to stop blackouts', *Sydney Morning Herald*, 23 March.

Vyas, C. and Gohn, B. (2012) *Executive Summary: Smart Grid Consumer Survey*, Pike Research, Boulder.

Warde, A. (2005) 'Consumption and theories of practice', *Journal of Consumer Culture*, vol. 5, no. 2: 131–53.

WEF (2009) *Accelerating Smart Grid Investments*, World Economic Forum (WEF) and Accenture Cologny.

Weiss, M., Mattern, F., Graml, T., Staake, T. and Fleisch, E. (2009) 'Handy feedback: Connecting smart meters with mobile phones', paper presented

to Proceedings of the 8th International Conference on Mobile and Ubiquitous Multimedia, Cambridge.
Wenger, E. (1998) *Communities of Practice: Learning, Meaning, and Identity*, Cambridge University Press, New York.
Wilhite, H. (2005) 'Why energy needs anthropology', *Anthropology Today*, vol. 21, no. 3: 1–2.
Wilhite, H. (2008a) *Consumption and the Transformation of Everyday Life: A View from South India*, Consumption and Everyday Life, Palgrave MacMillan, New York.
Wilhite, H. (2008b) 'New thinking on the agentive relationship between end-use technologies and energy-using practices', *Energy Efficiency*, vol. 1, no. 2: 121–30.
Wilhite, H. and Ling, R. (1995) 'Measured energy savings from a more informative energy bill', *Energy and Buildings*, vol. 22: 145–55.
Wilhite, H. and Lutzenhiser, L. (1999) 'Social loading and sustainable consumption', *Advances in consumer research*, vol. 26: 281–7.
Wilhite, H., Nakagami, H., Masuda, T., Yamaga, Y. and Haneda, H. (1996) 'A cross-cultural analysis of household energy use behaviour in Japan and Norway', *Energy Policy*, vol. 24, no. 9: 795–803.
Wilhite, H., Shove, E., Lutzenhiser, L. and Kempton, W. (2000) 'The legacy of twenty years of energy demand management: We know more about individual behaviour but next to nothing about demand', in E. Jochem, J. Sathaya and D. Bouille (eds), *Society, Behaviour and Climate Change Mitigation*, Kluwer Academic Publishers, Dordrecht, 109–26.
Wilk, R.R. and Wilhite, H. (1985) 'Why don't people weatherize their homes? An ethnographic solution', *Energy*, vol. 10, no. 5: 621–9.
Willum, O. (2008) Residential ICT Related Energy Consumption which is not Registered at the Electric Meters in the Residences, Willum Consult and DTU Management Engineering, Copenhagen.
Wimberly, J. (2011) *EcoPinion Consumer Cents for Smart Grid Survey Report, Issue 12*, EcoAlign, Washington, May, http://www.ecoalign.com>.
Winner, L. (1977) Autonomous Technology: Technics-out-of-Control as a Theme in Political Thought, The MIT Press, Cambridge, Massachusetts.
Wolsink, M. (2012) 'The research agenda on social acceptance of distributed generation in smart grids: Renewable as common pool resources', *Renewable and Sustainable Energy Reviews*, vol. 16, no. 1: 822–35.
Wood, G. and Newborough, M. (2003) 'Dynamic energy-consumption indicators for domestic appliances: Environment, behaviour and design', *Energy and Buildings*, vol. 35: 821–41.
Wood, G. and Newborough, M. (2007) 'Energy-use information transfer for intelligent homes: Enabling energy conservation with central and local displays', *Energy and Buildings*, vol. 39, no. 4: 495–503.
Woodruff, A., Augustin, S. and Foucault, B. (2007) 'Sabbath day home automation: "it's like mixing technology and religion"', paper presented to Proceedings of the SIGCHI conference on Human factors in computing systems, San Jose.

Woodruff, A., Hasbrouck, J. and Augustin, S. (2008) 'A bright green perspective on sustainable choices', in CHI 2008, ACM, Florence, 5–10 April, 313–22.

Wyche, S., Sengers, P. and Grinter, R. (2006) 'Historical analysis: Using the past to design the future', in P. Dourish and A. Friday (eds), *UbiComp 2006*, Springer-Verlag, Berlin/Heidelberg, vol. 4206: 35–51.

Zpryme 2011, *The New Energy Consumer*, Zpryme Smart Grid Insights, Austin.

Zwick, D. and Denegri Knott, J. (2009) 'Manufacturing Customers: The database as new means of production', *Journal of Consumer Culture*, vol. 9, no. 2: 221–47.

Index

Accenture, 39–40
active consumers, 31–2, 43–4, 137–9
actor-network theory, 62
Adaptive Control of Home Environment (ACHE), 130
agency, 25, 62, 63
air-conditioning, 21–2, 81, 85–6, 95, 96, 100, 101, 106, 109–14, 118–19, 121–2, *see also* comfort, cooling
American Society for Heating, Refrigeration and Air-Conditioning Engineers (ASHRAE), 120
appliances, 11–12, 21, 26, 47, 64, 78, 86–7, 107, 123–5
artefacts, 58
Association of Home Appliance Manufacturers (AHAM), 123
Australia, 20, 96, 105, 111
Australian Energy Market Commission (AEMC), 106–7
automation technologies *see* home automation technologies
Average Man, 38
Attitudes-Behaviour-Choice (ABC) model, 158

BC Hydro, 44–5
behavioural economics, 99
behavioural theories, 55–6
Berst, Jesse, 118
blackouts, 101–2, 108, 111
bodily performance, 57
Bourdieu, Pierre, 57brownouts, 111
Brattle Group (US), 110
California building codes, 122
California energy crisis, 102, 105, 110

Canada, 45
children, 82–3
citizen-consumer, 38–9
cold spots, 29, 107, 108, 113, 115, 128, 148
comfort studies, 112
commodification, 77
computer science, 38
congestion charging, 159–60
consumer behaviour, prices and, 99–101
consumer decision making, 10, 28, 32, 37, 43–4, 48, 52, 56, 60–1, 84, 86, 98, 114, 139, 159
consumer electronics, 47
consumer energy experience chain, 44
consumerism, 47
consumer research, 39–42, 50
consumer segmentation, 39–42, 164–65
consumption, 124
consumption feedback, 9, *see also* energy feedback
control, 76–7, 117–18, 119, 126
Control4, 174
cooking, 10, 58, 64, 91
cooling, 6, 10, 58, 79, 81, 85–6, 95, 96, 106, 109–14, 118–19, 121–2, *see also* air-conditioning
coordinating practices, 125–7
critical peak pricing (CPP), 11, 94–115, 160–1, *see also* dynamic pricing
critical peak rebate (CPR), 97
cultural determinism, 25

Darby, Sarah, 80
data, 27–30, 65–6
data ownership, 18

day-to-day activities, *see* everyday practice
decarbonisation, 2, 153
De Certeau, Michel, 5
demand management, *see* energy demand
defamiliarization, 68
delegating practices, 127–31
digital ethnographers, 120
dinner routines, 10, *see also* cooking
direct load control (DLC), 9, 11, 31, 118–23, 127
disciplinary technologies, 77, 123
DLC, *see* direct load control (DLC)
do-it-yourself (DIY) practice, 131–3
domestic routines, 10, 12, 58, 61–2, 79–83, 86–7, 91–3, 107–9, *see also* routines
downtime, 126, 128
duality of structure, 56
'dumb' behaviour, 53–5
dynamic pricing, 65–6, 94–115, 122, 160–1, *see also* critical peak pricing, time-of-use pricing

EcoMeter, 104–5
Economic Man, 38
economists, 27–8, 38
efficiency, 23, 25
electricity, 2, *see also* energy
electricity consumption, 47, *see also* energy consumption
electricity cures, 140–1
electricity industry, 8
electricity pricing, 43, 65–6, 94–115
electric vehicles, 9, 46
Electronic Cottage, 21
Ellul, Jacques, 23–4, 32
embodied knowledge, 57–8
emergent technologies, 19
empowerment, 32, 42–6
end-user programmers, 119–20
energy, 7
 in everyday practice, 53–69
 as material, 63–6, 140–3
 meaning of, 60–2
 micro-generation of, 135–54
 in practice, 62–6
 relationship with, 141, 147–9
energy action, 74
energy consumers, 25, 28–33, 67
 differentiation of, 39–42
 engagement with, 48–51
energy consumption, 4, 77, 137
 control of, 37
 information on, 44–5
 production and, 64–5
 reducing, 10
 research on, 60–1
energy demand, 3, 5
 management of, 121, 123
energy efficiency, 2
energy feedback, 10–11, 65–6, 73–93
 delivery of, 74
 effectiveness of, 74, 77–8, 160–1
 in everyday practice, 81–92
 frequency of, 78
 promotion of, 76
 Resource Man and, 75–7, 81–4
 savings from, 78–81
 in Smart Utopia, 75–8
 social feedback, 84–6, 89, 92
energy information, 44–5
energy-making practices, 135–54
Energy Orb, 104–5
energy practices, 66–7
energy-saving practices, 10–11
energy savings, 74, 78–81
Energy Savings Trust (EST), 47
energy service companies (ESCOs), 47, 49
engineers, 27–8, 38
entertainment appliances, 26
entrepreneurial self, 38–9
European Commission, 20
everyday life, 3–4, 10–11, 22, 33, 38, 53, 59, 95, 101, 163, 165–7
 air-conditioning in, 109–14
 elasticity of, 113–114
 energy in, 60

feedback in, 160
ICTs in, 147–8
messiness of, 54–55, 101, 105–6
micro-generation in, 135
rhythms of, 107, 108, 161
smart technology and, 7, 25, 155–6, 159
study of, 5–6
everyday practice, 4–9, 13, 53–69, 80, 139–40, 164
automation of, 116–34
energy feedback in, 81–92
negotiating, 88–92
e-waste, 47
expert knowledge, 61

fallback effect, 77
Faruqui, Ahmad, 95–96
feedback, *see* energy feedback
feed-in tariffs, 137
fitness technology, 88
flat pricing, 95
folk physiological theories, 60–1, 132
Foucault, Michel, 77, 123
Fox-Penner, Peter, 34
freezing, 6, 90
future, imagining of the, 17–33
gamification, 9
Garmin fitness technology, 88
gender, 36, 38

General Electric (GE), 47
Get Energy Fit programme, 45–6
Giddens, Anthony, 56
Global Utility Consumer Surveys, 41, 43
Golem, 128–30
greenhouse gas emissions, 10
green renovations, 80
Grid 2030 (DOE), 19–20

habitus, 57
Hargreaves, Tom, 81–7
Hacking, Ian, 41
handmade energy, 145–6
health knowledge, 61

heating, 6, 58, 64, 80–1, 95, 110, 118–19
heatwaves, 101
Hobson, Kersty, 79
home area networks, 25
home automation technologies, 9, 11–12, 25, 26, 66, 116–34, 161–2home energy management (HEM), 42–5, 47, 75–6, 80
home grown energy, 145–6
Home of the Future Consortium, 31–2
homes of tomorrow, 25
homo economicus, 38
homo faber, 38
homo optionis, 38
human body, 57, 89–90
human-computer interaction (HCI), 4, 69, 78
human passivity, 31, 37, 137, *see also* passive consumer

IBM Institute for Business Value, 36, 43
ICT-enabled practices, 6
immaterial materials, 7
individualism, 4
information, 44–5, 77
information communication technologies (ICTs), 1, 10, 18, 22, 23–7, 147–8
information-deficit model, 4
information exchange, 28–9
in-home displays (IHDs), 25, 44, 45, 75–6, 80, 104–5
intermediaries, 64, 139

Kill-A-Watt, 45, 91
knowledge
  embodied, *see* embodied knowledge
  expert, 61

Latour, Bruno, 62
laundry, 10, 81, 86
lay epidemiology, 61

Lefebvre, Henry, 5
l'homme moyen, 38
lifestyle, *see* smart lifestyle
linear technological transfer, 4
literary prophecy, 27
living standards, 26
load factor, 10
local energy, 135–54
Lutron, 26

Maller, Cecily, 64, 141, 150–2
Marres, Nortje, 79, 80, 139
Massachusetts Institute of Technology (MIT), 31–2
materiality, 6–7, 9, 63–6, 140–3
media studies, 4
memory scrapbook, 68
men, 38, 120
methodology, 68, 164–165
Michael, Mike, 59
micro-generation, 7, 9, 12, 46, 135–54
micro-grids, 9
mobile applications, 44, 45, *see also* smart phone
More, Thomas, 17
Mozer, Michael, 130

Navigant Research, 42
neoliberalism, 38–9
new energy consumer, *see* Resource Man; smart energy consumer
New York blackouts, 101–2, 108
New Zealand, 45
Normal Man, 38

ontology of everyday practice, 3, 8, 53–69
ordinary objects, 59
Orthodox Jews, 127–9
outdoor heating, 80–1, *see also* heating

Pacific Gas and Electric Company (PG&E), 42
passive consumers, 37, 46, 49, 118–23, 137–9

peak electricity demand, 2, 9–10, 22, 96–7, 106–7
peak pricing, 11, 94–115, 160–1
peaky domestic appliances, 21–2
performativity, 7
Photovoltaic (PV) panels, 136, 140–4
Pickering, Andrew, 62
Pierce, James, 65, 78, 81–7, 141, 145–6, 150
Pink, Sarah, 60, 89
politics, 24–5, 122–3
PowerCentsDC Program, 105
Power Hog, 45
Power Meter, 45
practice-as-performance, 7
practice-as-performative, 7
practices, 5, 55–9
  coordinating, 125–7
  delegating, 127–31
  DIY, 131–3
  energy, 66–7
  energy-making, 135–54
  everyday. *see* everyday practice
  non-negotiable, 86–7, 92
practice theory, *see* social practice theories
prices
  dynamic, *see* dynamic pricing
  meaning of, 99–101
prices-to-devices, 31
privatisation, 18
programmable thermostat, 119, 132
prosumers, 48, 137
psychology, 99

rational choice theory, 4, 28–9, 32, 37, 137
real time pricing (RTP), 97–8
Reckwitz, Andreas, 63
regulation, 18, 24
relationship models, 48–51
reliability, 20
renewable energy, 2, 46
research techniques, 68–9
resource bias, 33
Resource Man, 2, 3, 8–10, 32–52

characteristics of, 36–7
concept of, 34–9
consumer research on, 39–42
empowering, 42–6
energy feedback and, 75, 76–7, 81–4
engagement with, 48–51
home automation technologies and, 120
smart lifestyle and, 46–8
world without, 156–9
ripple control systems, 119
routines, 106–9, 113–15, 125–7, see also domestic routines

Sabbath, 127–9, 130
Schatzki, Theodore, 5, 65
scholastic bias, 33
Schumacher, Ernst Friedrich, 161
science and technology studies (STS), 6, 7, 62, 63–4
'shared embodied know-how', 5
set-and-forget, 31, 37, 119
Shove, Elizabeth, 5–6, 57–9, 88
showering, 6, 81
slow technology, 148–9
'small is beautiful', 149, 161
smart appliances, 11–12, 21–2, 47, 107, 123–5
smart energy consumers, 1, 7–9, 25, 28–33
see also Resource Man
smart energy technologies, 2, 3, 7
research on, 8
roll-outs of, 18, 24
vision for, 17–33
Smart Grid Consumer Collaborative (SGCC), 30, 42
smart grids, 2, 7, 18, 20, 25, 32, 43
engagement and, 49
installation of, 22
research on, 8
smart homes, 21, 25–6, 42–3, 116–34
'smart' label, 1–2
smart lifestyle, 46–8, 123–4, 157

smart meters, 2, 7, 18, 20, 32, 43, 48
cost-benefit analysis of, 74
installation of, 22
research on, 8
smart ontology, 2, 3, 8, 18, 22–33, 56, 94, 126, 137, 162–5
smart phone, 108
smart strategies, 159–62
smart thermostats, 11, 118–20, 121–2, 126
Smart Utopia, 2, 3
data in, 27–30
energy feedback in, 73–93
home automation technologies in, 116–34
imagining, 17–33
technology in, 23–7
reimagining, 155–67
smart utopians, 13, 48
social action triggers, 45–6
social change, 19
social control, 24, see also control
social feedback, 84–6, 89, 92, see also energy feedback
social loading, 106
social networking, 88
social order, 19
social practices, 55–9
social practice theories, 3, 4–8, 55–9, 61–3, 156, see also practice theorysolar energy, 141–2, 143
Southerton, Dale, 107–8
Standard 55, 120
supply security, 103

technique, 23–5, 32, see also Ellul, Jacques
technological substitution, 24
technological utopias, 17, 22, 26
technology, 23–7
temporality, 147–50
Theory of Interpersonal Behaviour, 87
thermal comfort, see comfort studies

thermostats, 118–22, 126, *see also* programmable thermostats, smart thermostats
The Third Wave (Toffler), 21
time-based pricing, 9, *see also* time-of-use (TOU) tariffs
time constraints, 107–8, 126–7
time-of-use (TOU) tariffs, 97, 98, *see also* dynamic price; time-based pricing
time-space dynamics, 147–8
Toffler, Alvin, 21
Tool Man, 38
tweetawatt, 45

ubiquitous computing, 21, 55
United Kingdom, 20, 45, 47, 48, 81, 107, 109, 143, 144, 146

United States, 21, 42, 45, 50, 96, 101, 105, 109, 119, 132
US Department of Energy (DOE), 19–20
user-centred design, 4
Utopia (More), 17

variable pricing, *see* dynamic pricing

weatherisation, 61
web-based portals, 44–5
Westinghouse, 25
wind energy, 141–2
women, 38, 82–3, 120
Woodruff, Allison, 128–30

Zpryme Smart Grid Insights, 36

Printed and bound by CPI Group (UK) Ltd, Croydon, CR0 4YY

The manufacturer's authorised representative in the EU is Springer Nature Customer Service Centre GmbH, Europaplatz 3, 69115 Heidelberg, Germany. If you have any concerns regarding our products, please contact ProductSafety@springernature.com

Printed and bound by CPI Group (UK) Ltd, Croydon, CR0 4YY

23/03/2026

02076449-0008